Sandscapes

Jo Carruthers · Nour Dakkak
Editors

Sandscapes

Writing the British Seaside

palgrave
macmillan

Editors
Jo Carruthers
Lancaster University
Lancaster, UK

Nour Dakkak
Arab Open University
Ardiya, Kuwait

ISBN 978-3-030-44779-3   ISBN 978-3-030-44780-9   (eBook)
https://doi.org/10.1007/978-3-030-44780-9

Cover credit: MartiniDry, shutterstock.com

This Palgrave Macmillan imprint is published by the registered company Springer Nature Switzerland AG
The registered company address is: Gewerbestrasse 11, 6330 Cham, Switzerland

Acknowledgements

The contributors to this book met over two days in a warm but windy More-cambe in April 2017. We walked and wrote on the sands, talked about sand mining, golf courses, and—inevitably—the weather. That set of weekend meetings was funded by the Arts and Social Sciences Faculty Research Fund at Lancaster University, and we thank them for enabling that important chance to explore Morecambe and for us to share our first thoughts and drafts. Thanks also go to the wonderful Morecambe Hotel who generously let us use their meeting room and provided us with a gorgeous lunch as well as to The Beach Café for their flexibility for our lunchtime meeting as we took over much of the café one blustery Friday in April.

As Editors, we are extremely grateful to each of the wonderful contributors to this volume, who have given their time, energy and creativity while working within increasingly stretched and pressured Arts, Humanities, Social Sciences and Higher Education sectors. We are very aware that our jobs were made easier by the contributors' efficiency and hard work. Thanks are also due to the editors at Palgrave, Allie Troyanos and Rachel Jacobe, who have been unremittingly efficient, helpful and an absolute joy to work with. We would also like to thank the anonymous reviewers of the proposal and full manuscript who generously offered suggestions for further reading, thoughts and revisions. We are grateful for our colleagues in the English Literature and Creative Writing Department at Lancaster and notably those wonderful people who read material for us beyond the call of duty, including Jenn Ashworth, Michael Greaney, Sharon Ruston, Rebecca Spence, Catherine

Spooner, Andrew Tate and Eoghan Walls. Thank you also to friends and family who were generous sounding boards: Helen Wilkinson—and Tom, Patrick, Isabel and Edward—Sue Jackson, Joud, Richard, Molly and Elliot.

Contents

Notes on Contributors

Jenn Ashworth is a Senior Lecturer in Creative Writing at Lancaster University. She has published four novels (her latest, *Fell*, is set on the North West coast of England) and one book of creative-critical essays: *Notes Made While Falling* (Goldsmiths, 2019). She writes short fiction for broadcast and various publications and was the editor of *Seaside Special: Postcards from the Edge*, an anthology of new writing about the North West Coast. More at www.jennashworth.co.uk.

Brian Baker is a Senior Lecturer in English and Creative Writing at Lancaster University, UK. He has published books on masculinities and on science fiction, including *Contemporary Masculinities in Fiction, Film and Television* (Bloomsbury Academic, 2015) and *The Reader's Guide to Essential Criticism: Science Fiction* (Palgrave Macmillan, 2014). He has recently written on Alan Garner, on masculinity and class in *Quadrophenia*, and on *Tinker Tailor Soldier Spy*. He is currently engaged in several critical and creative projects that investigate concepts of collage and transmission across literature and media and is taking a Master's degree in Art Practice.

Julian Brigstocke lectures in Human Geography at Cardiff University. His research focuses on the geohumanities, urban cultures, geographies of authority, and speculative and materialist theory. He is author of *The Life of the City: Space, Humour, and the Experience of Truth in Fin-de-siècle Montmartre* (Ashgate, 2014) and co-editor of *Space, Power and the Commons* (Routledge, 2015) and *Listening With Non-human Others* (ARN Press, 2016).

Jo Carruthers is Senior Lecturer in English Literature at Lancaster University and researches in the areas of literary studies, aesthetics and religion. Her publications include *Anticipatory Materialisms in Literature and Philosophy, 1790–1930* (Palgrave, 2020), edited with Nour Dakkak and Rebecca Spence, and a history of simplicity in *England's Secular Scripture: Islamophobia and the Protestant Aesthetic* (Continuum, 2011) and "Aesthetics of Simplicity" in the 2019 *Routledge Companion to Literature and Religion* (2016). She is currently working on a book on the aesthetics of roughness and its relation to class in Victorian fiction.

Peter Coates is Professor of American and Environmental History at the University of Bristol, UK, and currently works on animals (including mosquitoes, eels and squirrels), watery environments (salty and fresh), wildlife filmmaking and notions of bio-cultural heritage. Recent books include *A Story of Six Rivers: History, Ecology and Culture* (Reaktion, 2013) and the co-edited (with David Moon and Paul Warde), *Local Places, Global Processes: Histories of Environmental Change in Britain and Beyond* (Windgather/Oxbow, 2016). His landscape of childhood was the broad sandy beach and dunes at Formby, Merseyside.

Tim Cole is a historian who is currently completing a book for Bloomsbury non-fiction trade called *About Britain*, that retraces itineraries created for the Festival of Britain seventy years later. His previous books have focused on the landscapes and memory of the Holocaust—*Images of the Holocaust/Selling the Holocaust* (1999), *Holocaust City* (2003), *Traces of the Holocaust* (2009) and *Holocaust Landscapes* (2016). Tim is Professor of Social History at the University of Bristol, where he is also Director of the Brigstow Institute.

David Cooper is a Senior Lecturer in English at Manchester Metropolitan University, UK, where he co-directs the Centre for Place Writing. A founding co-editor of the interdisciplinary journal, *Literary Geographies*, he has published widely on contemporary British place writing. He has also published extensively on digital literary mapping and is currently Co-Investigator on "Chronotopic Cartographies" (Arts and Humanities Research Council). David is increasingly experimenting with creative-critical approaches to place and collaborated with the artists, Michelle Green, Caro C, and Maya Chowdhury, on "Hayling Island: Stories at Sea Level": a multi-media mapping project funded by Arts Council England.

Nour Dakkak is Assistant Professor at the Arab Open University, Kuwait where she teaches humanities and literature. Her publications include *Anticipatory Materialisms in Literature and Philosophy, 1790–1930* (Palgrave, 2020), co-edited with Jo Carruthers and Rebecca Spence, and "Mobility, Attentiveness, and Sympathy in E. M. Forster's Fiction" in *Mobilities, Literature,*

Culture, edited by Marian Aguiar, Charlotte Mathieson and Lynne Pearce (Palgrave, 2019). She is currently working on a book on material humanism in the writings of E. M. Forster.

Christopher Donaldson is Lecturer in Cultural History at Lancaster University, UK, where he is also Research Centre Coordinator of The Ruskin—Library, Museum and Research Centre and an associate of the Regional Heritage Centre. His research is principally concerned with changing perceptions of the value of landscape and the environment during the eighteenth and nineteenth centuries and with the way these perceptions were mediated by commercially produced guidebooks and topographical art and literature.

Michelle Green is a UK-based writer and artist. Their debut collection of short fiction—*Jebel Marra* (Comma Press, 2015)—was supported by Arts Council England and the Julia Darling Fellowship, and was nominated for the Polari First Book Prize, the Edge Hill Prize, and the Frank O'Connor International Short Story Award. Their ongoing work on disability aesthetics was featured at the Mathrubhumi International Festival of Letters 2019, and they are now working on a second collection: a map of short stories based on Hayling Island. More at www.michellegreen.co.uk.

Shona Legaspi has worked on various oral history projects, including recollections of the journeys Polish migrants undertook to the UK during and after World War II. She is currently developing a project to further research the history of women in Morecambe after becoming inspired during a Ph.D. started in the Sociology department at Lancaster University.

Sefryn Penrose is an archaeologist who specialises in developing new approaches to heritage. Much of her work has focused on post-World War II Britain, developing a better understanding of how everyday change and creation across the later twentieth century has affected the landscape, and the way we have lived it. She is the author of *Images of Change: An Archaeology of England's Contemporary Landscape*, published by English Heritage. She is the sort of person that swims at every opportunity and once swam the non-tidal Thames on weekends over two years.

Angela Piccini is a Reader in Screen Media at Bristol University and is one half of Bureau of the Contemporary and Historic (ButCH). As an artist and educator, she explores lively materialities through film and video with a particular focus on heritage-based creative (non)fictions. Her work includes *Guttersnipe—A Micro Road Movie* (2004), *Beachley-Aust* (2009), *Association of Unknown Shores* (2018-date, with Kayle Brandon and others) and she has published *Contemporary Archaeologies: Excavating Now* (2009, co-edited with Cornelius Holtorf), *The Oxford Handbook of the Archaeology of the Contemporary World* (2013, co-edited with Paul Graves-Brown and Rodney Harrison), and *Imagining Regulation Differently: Co-creating Regulation for Engagement* (2020, co-edited with Morag McDermont and Tim Cole and Janet Newman). She grew up in Vancouver, swimming at urban beaches and feeding mussels to the sea anemones.

Jean Sprackland is the author of *Strands: A Year of Discoveries on the Beach*, the winner of the Portico Prize for Non-Fiction in 2012, and of *These Silent Mansions: A Life in Graveyards* (2020). She has also published five collections of poetry, most recently *Green Noise* (2019), and she won the Costa Poetry Award in 2007 with *Tilt*. Jean is Professor of Creative Writing at Manchester Metropolitan University.

List of Figures

1

Introduction: Sandscapes

Jo Carruthers and Nour Dakkak

The Sand at Our Feet

On a three-kilometre stretch of beach at Crosby, near Liverpool, stand one hundred cast-iron naked men. Replicas made by the artist Antony Gormley of his own body, the statues—still and uncanny—face out to the cold Atlantic Ocean. Any tourist consulting the "Visit Liverpool" website will learn that each figure weighs 650 kilos and together they look over the water "in silent expectation". Some of the figures are submerged in the sea, some at a distance dangerous for flesh-and-blood at nearly a kilometre from the promenade.[1] The figures represent strange points of stability, reliant on a three-metre foundation pile in order to stand erect on the otherwise shifting—and often dangerous—sands. These figures—marked, rusted, corroding and buried to different depths in the sandscape—epitomise the hypnotic draw of the sea, as their (iron) eyes look away from the sand in which they stand. The installation is called "Another Place", perhaps alluding to that other world of the horizon, but just as likely to point towards the alien world of the sandscape to which tourists' gazes are drawn by the crowd of sculptural Gormleys. These figures are fascinating not only for where they look, but for their position in the sand, for how they are weathered, buried, covered and battered by it.

J. Carruthers (✉)
Lancaster University, Lancaster, UK

N. Dakkak
Arab Open University, Ardiya, Kuwait

© The Author(s) 2020
J. Carruthers and N. Dakkak (eds.), *Sandscapes*,
https://doi.org/10.1007/978-3-030-44780-9_1

The writers of the essays collected in *Sandscapes* resist the urge to look out at the horizon between sea and sky and instead turn to sand itself. Popular, creative and critical studies of the seaside alike have made more of the sea and the horizon—and even the air, pebbles, rocks or cliffs—than of the sand.[2] This book considers the sandy edge of land that is more commonly called a beach, shore, coast or seaside. These familiar terms tell us how sandscapes are used and what they have come to represent: a space to play on, the boundary to the sea, a holiday destination. *Sandscapes* focuses on how sand as terrain—soft white sandy beaches, hard yellow foreshores, quicksand, muddy estuaries or sand dunes—exists in our imaginations and in British culture at large. Sand has an important status in British culture especially: everyone is not much more than an hour's drive from the shore. The unusual physical qualities of sand—malleability, instability, portability—have become the stuff of legend and of aphorism. In this book, writers have responded to the sandscapes they have visited, read about or imagined. They draw on sand's playfulness, threat and mobility to think through the environmental, social, personal, cultural and political status of sand and the seaside towns that have built up around it. Indeed, the range in style and focus reflects the granular and playful qualities of sand itself, to express the unique topography of sand, sandscapes and the British seaside town in their material, literary, representational, sociological and figurative richness.

The Seaside Town

The seaside town holds a contradictory position in the British cultural imagination: the epitome of post-war nostalgia, the seaside is summed up by the ubiquitous family snapshot of children making sandcastles. The home of cheeky postcards, sands may also be visited alone for moments of silent meditation of the kind that Gormley's sculptures embody. More recently, seaside towns have made the news as the epitome of austerity-driven deprivation. They were earlier brought to national attention in the case of the Chinese migrants killed in the Morecambe Bay cockling disaster of 2004, grimly dramatised in Nick Broomfield's film, *Ghosts*.[3] Sandscapes are places of work—sometimes dangerous work—as much as play. The topographic specificity of this in-between place—neither seascape nor landscape—has too often been overlooked, despite sand playing such an engaging part in the national and global imaginary.

As the edges of nations, beaches were for a long time places of invasion or departure for war, many seaside towns growing up alongside military bases, visitors often drawn by their regimental bands.[4] Before the invention of more efficient land transportation in the nineteenth century, sandscapes would have more often been points of departure and arrival, as explored in Christopher Donaldson's chapter on the crossing of sands to the Lake District. Although people were visiting the coast for respite from the seventeenth century onward, it was only in the Victorian period that the seaside became associated with the family holiday. In the nineteenth century, people flocked to beaches because of the health-giving properties of the sea and sea air. Historian of the seaside, John K. Walton, dates the growth of seaside resorts to the co-emergence of the train and wage industries that produced surplus incomes to be spent on leisure.[5] Here, the iconic British seaside was born, as Lewis Carroll's *Alice in Wonderland* testifies:

> Alice had been to the seaside once in her life, and had come to the general conclusion that, wherever you go to on the English coast you find a number of bathing-machines in the sea, some children digging in the sand with wooden spades, then a row of lodging-houses, and behind them a railway-station.[6]

In academic studies, attention given to the health benefits of sea-bathing and sea air often eclipse activities that take place on the sandscape. Yet the sandy foreshore that skirts Britain was vital to the evolution of seaside resorts, including the social disruption caused by the mix of people walking, horse riding or even driving on the sands.[7] In his sociological studies of seaside towns, Walton draws attention to the issues of class integration that informed the specific growth and make-up of seaside towns from their very beginnings. In the eighteenth and nineteenth centuries, sands were places where the full range of society could mix, a muddle of people that disrupted inscribed social hierarchies. Developments in the railway network and increased incomes eventually led to longer holidays for weavers and factory workers, who would decamp on mass during the summer factory shutdown. We can see the potential freedoms and licentiousness of the seaside dramatised in Stanley Houghton's play, *Hindle Wakes* (1910), later a film directed by Arthur Crabtree (1952).

In the nineteenth century, seaside towns would be flooded with visitors looking for a home away from home. Resorts catered for the lower and higher classes, providing different kinds and types of activities and different sandscapes within them. Andrew Davies's 2019 adaptation of Jane Austen's unfinished and final novel, *Sanditon* (1817), populates a South East seaside resort (probably inspired by Bognor Regis) with not only naked swimming by men, but also "bathing machines" that took women out to deeper waters. Women would enter these "mobile chariots", as early seaside historian H. G.

Stokes describes them, to change into modest suits that offered full coverage and be transported to the deeper waters that would protect their dignity. They would jump into the sea to be helped by the women who operated the machines, waiting in the water to assist them.[8]

Sandscapes remained ambivalent spaces into the twentieth century. The beach became a strange icon of World War II bravery, with Dunkirk and Winston Churchill declaring to a nation at war: "we will fight them on the beaches". Tourist guides to beaches sprung up alongside the motorcar in the early twentieth century and footfall increased to the seaside resorts. As Tim Cole discusses in his chapter, this new mobility afforded access to and greater appreciation of coastal scenery rather than tourists being limited to resorts served by railway stations. Celebrations of the seaside often return to this point in time—to buckets and spades, sandy sandwiches, naughty post-cards—when the home-grown holiday, necessitated by post-war economics, was the norm. Yet beyond this nostalgic gloss and the upmarket celebration of the simple life in places like Margate, the seaside town has become notorious as a site of decline. Michael Bracewell and Linder, using Morecambe as their focus, describe the "vivid contrast" of the "now deliquescent resort" which contains on the one hand "near dereliction and social problems" and on the other the "enduring gentility, grace and exotica of a once thriving town".[9] Topographically and socially marginalised, seaside towns have nonetheless proved hospitable to organic, community and cultural projects that tap into the natural and cultural resources of this distinctive habitat.

Sand may be imaginatively provocative and the sandscape poetically fertile, but it has for a long time been a matter of politics. Thinking about seaside life purely in terms of leisure and entertainment overlooks critical implications about class, race and minority groups residing and living in these places, as Daniel Burdsey argues in his book, *Race, Place and the Seaside*. Burdsey turns to the sea-inspired metaphor of "coastal liquidity" to defy nostalgic or static notions of coastal identities and especially to challenge representations of the seaside town as a white space.[10] Sand—with its granular, shifting and congealing morphology—is a fitting image through which to approach the seaside town. Sand can help us to think about seaside identities—often so tied up with national myths, nostalgic memories and cultural symbols—as formations that intermittently coagulate or soften as well as shift. Mythologies, like sand, are subject to moments of positive as well as negative solidification but are reassuringly open to change.

Seaside towns and resorts are produced by and through the continuous movements of people in them, through tourism, holidaymaking, working. Recent academic studies of tourist practices focus precisely on the formative processes of such practices: that places are dynamic environments because what people do in them changes over time. David Crouch has argued that, through engaging in different types of process, people become producers rather than just consumers of leisure, and "space ceases to be only objective, contextual and metaphorical. Place becomes the material of popular culture which is worked, reworked and negotiated", primarily through human activities within a certain social and material environment.[11] People act upon sand, but sand also acts upon people. While sand itself is often overlooked when considering the seaside, the chapters in this collection—precisely because they are investigations of sandscapes—aim to understand the seaside as a dynamic space. This volume undermines insular or fixed readings of the coast, the beach or the shore. Turning to sand itself, the chapters to follow aim to abolish rigid boundaries—between the human and the nonhuman and between disciplinary borders—by bringing together normally separated critical and creative practices in an attempt to explore and reconfigure the British seaside.

A Brief Cultural History of Sand

A symbol of aridity, sand has nonetheless been a fertile site for cultural meaning. Sandscapes have a rich cultural history that provides a way into understanding the specificities of British seaside towns and shores. Familiar and yet so infrequently dwelt upon, the sandscape is a recurrent feature in literature, film, music and art. Sand's cultural richness emerges in part from its distinct physical properties. Sand's granular networks make it highly porous and liquefied—add enough water and it can be poured from a bucket or through hands. When added to cement, the coarse framework produces a rock-like strength. Technical books on sand read like poetic appeals to its beauty. Testifying to sand's drawing together of the seemingly opposed worlds of science and art, they speak of its "framework of grains", its "interlocking crystalline mosaic with zero porosity".[12] It is precisely this granular framework that has been so extraordinarily generative, the key to its abundance, malleability and variability. Its peculiar physical properties have made it a fundamental ingredient for the construction industry: this is the gritty substance of the high-rise and land reclamation. Vince Beiser's book, *The World in a Grain*, explains how sand has and continues to transform human civilisation.[13] Sand is more crucial to the massive industries and projects which shape the face of our globe than we might at first imagine. Its practical and political import is tied to a shortage in specific grades of sand and makes it a focus for concern in a world of increasing ecological precarity. As Julian Brigstocke's chapter in this volume makes clear, how we "think with" and imagine sand is integral to our attitudes towards it and may well be promissory for more sustainable futures.

Sand is like no other landscape substance. It moves, disperses, but it can also become dense and compact to the point of seeming solidity. Sand is chaotic, as Steven Connor eloquently explains:

> Sand belongs to the great, diffuse class, undeclared, rarely described, but insistent and insinuating, of what may be called quasi-choate matters – among them mist, smoke, dust, snow, sugar, cinders, sleet, soap, syrup, mud, toffee, grit. Such pseudo-substances hover, drift and ooze between consistency and dissolution, holding together even as they come apart from themselves. And, of all of these dishesive matters, sand is surely the most untrustworthy, the most shifting and shifty.[14]

Sand disperses and solidifies. It is absorptive; it makes visible the weather when it is blown into the air through which we walk, and when it moves in response to the tides. Our aim in this book is to show how attending to sand in all of its dynamic and substantial vitality is crucial to British seaside writing. We explore how sand is integrated in the politics and poetics of the environment not by attending to its practical functions, but to its qualities as a dynamic matter with its own independent ontology; in other words, to what makes sand sandy. This has meant involving writers from a spectrum of disciplines—history, geography, heritage, screen and literature studies, novelists and poets—to speculate on sand's aesthetics, uses, values and histories. The sandscapes presented in this volume are not exhausted by their relation only to humans or how they have been perceived by humans. After all, sands are the home of the crab as much as to us. Instead, the writings collected here all testify to the fact that all things and objects—all matter—is vibrant, as Jane Bennett has argued. Humans, nonhumans, objects and things alike have their own purposes and potentials.[15]

Bennett's concept of vibrant matter compels us to diminish perceived boundaries between subjects and objects, between "us" and "them", or "us" and "it". Her book argues for the reduction of humanity's role in the making and shaping of experience and demands that we acknowledge nonhuman activity in the development of humans and societies. As we know all too well, sand gets into places it is not welcome. It is very often beyond our control. As such, *Sandscapes* as a collection aims to show how the peculiar materiality of sand is closely integrated with the activities associated with the seaside. As new scientific studies increasingly reveal, seemingly inert objects such as sand are strangely active, and sand acts as a catalyst in the many relations, actions and developments of the societies, memories, identities and imaginations of seaside towns. Sand is not just linked but enmeshed with human bodies, just like Gormley's human figures are embedded within Crosby's sandscape.

Augurs of grand philosophical truths of time and eternity, sandscapes close-up are messy places. Thin lines, strands, intermittently edging the coastline, sandscapes are also grainy, muddy, grassy and inhabited. Smooth, solid and calm from a distance, up close sand is rough, malleable, teeming with life. The children's Ladybird guide, *The Seashore and Seashore Life*, focuses on sand as habitat and home.[16] Through its finely drawn illustrations, young readers can learn about the insects and animals that they may observe at the beach. Seemingly barren, sand is revealed here as the home of the lugworm or the periwinkle. This guide leads you on a virtual adventure as it tells, "Just look", and asks "Can you see...?", "Watch", and even "But stop!" As Elizabeth Bishop's poem "The Sandpiper" reminds us, the sandscape belongs more to birds and insects than to humans.

This book charts, then, the entangling of material and cultural sandscapes; the physical properties of sand from which its imagined life depends. Sand—as an object held in the hand or seen from afar along the coast—speaks to us of holidays and respite, but also of time and mortality, of plenitude and eternity. It speaks of mobility, movement and transience. Sand has long been identified with the infinite. In his poem "Auguries of Innocence", William Blake compels us "To see a World in a Grain of Sand".[17] Kahlil Gibran writes that humans and the world "are but a grain of sand upon the infinite shore of an infinite sea".[18] Both writers invoke the eternal alongside the finite, Blake to express the profundity of nature's everyday miracles, Gibran emphasising creaturely humility. In Jorge Luis Borges's short story, "The Book of Sand", sand's seeming infinity is embodied in a book with seemingly endless pages and shifting inscriptions. The book's infinitude amazes but also frustrates its owner because it represents a dangerous lack of boundary. Sand in Borges's story signifies the impossibility of knowing *for sure*, its defiance of human control unsettling knowledge itself. It is perhaps no surprise that the infinity that sand represents also provokes uncertainty in its exposure of human limitations. Sand in its capaciousness and its waywardness is all too often beyond our control and—as the owner of the "Book of Sand" eventually realises—it bespeaks mortality, finitude and our own limitations. In the Bible's Book of Genesis (22.17), God promises the childless Abraham that he will have descendants as numerous as the sands on the seashore, demonstrating a divine munificence that both makes up for and exposes human frailty.

Jesus writes on the sand when people bring a "woman caught in adultery" to him in the Gospel of John, Chapter 8. Faced with the merciless judgement of a group of men, Jesus chooses the impermanence and transience of sand with which to write something that averts his audience's judgment. Again, divine benevolence and human shortcomings coincide. Sand somehow represents to us human frailty and mortality encompassed in the slow flow of sand through an hourglass, a symbol found in early modern *vanitas* paintings, on gravestones and less poignantly through the opening credits to the US TV series *Days of Our Lives* (1965–) in which the aphorism "Like sands through the hourglass, so are the days of our lives" is played over a soundtrack that morphs from an ominous ticking of a clock to a sweeping romantic melody. One of the most provocative reminders of the sandy ravages of time is Percy Bysshe Shelley's poem "Ozymandias", a stark image of hubris laid waste. It is sand's invocation of temporality that E. M. Forster makes use of in his novel *Maurice*, as explored by Nour Dakkak in her chapter.

In the nineteenth century, we find the great "men of letters" and many notable women practicing their scientific endeavours on the seashore, drawn to the sandscape for its teeming life. Charles Kingsley, author of *The Water Babies*, was also aquatically inspired to pen his strange exploration of seaside life, *Glaucus: Or the Wonders of the Shore* (1856). George Henry Lewes and George Eliot escaped to the sandscape (a space infamous even then for unmarried liaisons) in order to write Lewes's *Seashore Studies* (1858), with Eliot leaving behind notes on her trips to Ilfracombe and Tenby, published in her collected letters.[19] The sands of Lyme Regis—made famous by Austen's *Persuasion* in 1817—was also the site at which Mary Anning discovered the first Ichthyosaur skeleton in 1811, testifying to the sandscape's function as geological archive. In William Dyce's painting, *Pegwell Bay, Kent—A Recollection of October 5th 1858*, the artist reproduces his family collecting shells but also paints the sandscape as a place that signifies cosmological time. A comet can be seen faintly in the daytime sky and the cliffs that face the beach display geological strata revealing a history predating human habitation. The red of the women's clothes reflect the hue of the sunset, linking the quotidian present to more cosmic timescales.

Caught on the wind, adhering to skin, clothes and food, sand is perhaps best known for its disregard of boundaries. While sand at the beach provides a background to bodies that bathe, walk and play, its positive associations tend to be modified as soon as it fails to fulfil its function as a soft support for our bodies and becomes an intruder instead. Or worse, when it threatens to overpower. Writers and artists have turned to the instabilities of sand—how sands move by the day, the hour, and even the minute—to explain some-thing about their own mutable experiences. Writers in this volume—notably Jenn Ashworth, Brian Baker, Jo Carruthers and Jean Sprackland—find the sandscape a place of quiet, sometimes desperate, meditation. Sand acts as a metaphor for life: shifting under the caress or violence of the sea, marks are made and then erased, covering or unearthing things buried, to tell stories of chaos, freedom and even danger.

Sand's instability often becomes a metaphor for life itself. In Jenn Ashworth's 2016 novel, *Fell*, uncertainty within and between the characters, real and ghostly, replicates the shifting terrain of Grange-over-sands: "The gullies and channels shift, the sands run like mercury: no one can trace the same path across them twice".[20] In Matthew Arnold's "Dover Beach", the tide's ebb and flow on the shore represents nineteenth-century Britain's (and Arnold's own) increasing rejection of religious faith. In the poem, the sea retracts from the sandscape like the withdrawal of the infinite—figured by the "sea of faith" (line 21)—so that the speaker stands, but not so firmly, on the disenchanted "naked shingles of the world" (line 28).[21] It is the very in-betweenness of the sands—they have been utterly shaped by and are still reliant on, the movement of the "Sea of Faith"—that sums up Arnold's torn attitude towards this ebbing away of religious belief. This unsettling becomes a certain kind of haunted disturbance in Andrew Michael Hurley's novel, *The Loney* (2014), set in a thinly disguised Morecambe Bay. Father Bernard, an anchor of faithful stability in the novel, asserts God's control in a messy world. The Father repeatedly encourages his parishioners that God is still present even through suffering. He goes to the sand specifically to find God: "Here was the wild God, who made nature heave and bellow", but instead finds himself bereft.[22] Sandscapes are not always affirming places.

The association between sand and danger perhaps explains why sandscapes are so hospitable to imagined horrors. The "Shivering Sand" is a mysterious and terrifying place in the classic sensation novel, *The Moonstone* (1868) by Wilkie Collins. The "shivering and trembling" quicksand allures the servant Rosanna, who observes ominously to the narrator Mr. Betteredge: "I think that my grave is waiting for me here".[23] In the edgeland of sand and sea in Sarah Perry's *The Essex Serpent* (2016) there exist real and imagined monsters.[24] Less disturbing is the monster of E. Nesbit's *Five Children and It* (1902). Deprived of a real seaside, the children in this story take their "Margate spades" and dig in gravel-pits. "We can pretend it's seaside" they say.[25] There they dig up a "sand fairy", an unexpected finding of a peculiar kind of treasure in their imitation of seaside play.

The sandscape is a site that epitomises childhood in bucket-and-spade play. Adults also play here—uncharacteristically—so that normally dignified adults take off their shoes to wade in crashing waves. The seaside is where people go to have fun precisely because it is a place where social niceties can be left behind. A place of pleasure—as the collection *Modernism on Sea* has argued—the sandscape is inevitably also a place of risk, danger and disruption.[26] The pleasures of the seaside take it outside of everyday societal norms: it is the place for honeymoons, secretive trysts and scandals. Even in ancient times, the sandscape had an erotic mythology for Nausicaa, host to the newly-arrived, naked Odysseus in Homer's epic, now captured in the Eric Gill bas-relief in Portland stone in the Midland Hotel, Morecambe. A disastrous chastity is experienced by Florence and Edward in Ian McEwan's *On Chesil Beach* (2007), the cold pebbles metonymically replicating their failed intimacy.[27] (One wonders if everything would have been different if they had been lazing on yielding sands, bare feet exposed to warm grains, the sandscape drawing out an otherwise suppressed playfulness. In the film version, they are fully clothed and wear shoes on inhospitable shingles.)

The beaches in the much sandier *Weymouth Sands* (1934) by John Cowper Powys are a site of various disreputable liaisons (so scandalous it seems that the novel was first published under the title *Jobber Skalds*, after one of its main characters, because the town's corporation objected to Weymouth being associated with such licentious behaviour). As the novel flits from character to character, offering differing perspectives on the Weymouth community, much of the action happens on the sands, so near yet so different to Chesil Beach. The sands become a world in microcosm—what John Bayley calls a "swirling vortex"—that fizzles with danger for lovers, mystics and philosophers alike.[28] Influenced by Thomas Hardy, Powys presents a novel in which life is sensitive to its environment: the holy fool who preaches on the sand, the Punch and Judy performer, workers leading itinerant and unsettled lives; the family at the local inn, with its too-knowing children, who rent out rooms to visiting lovers. Powys's writing invokes D. H. Lawrence's erotically charged novels and like them presents sexuality as inchoate as sands. Bayley also calls the novel's style "peristaltic" referring to the wave-like movements of the body's natural processes—the muscular movement that pushes food through the body—so that Powys's fragmented narratives resemble the gentle agitations of the sandscape. It is a novel preoccupied with the grainy intricacies of a heady mix of the mundane, freedom and danger.

Sandscapes often signal a vaguer sense of misdemeanour. When Lydia Bennett runs away with Mr. Wickham in Austen's *Pride and Prejudice* (1813), they go to Brighton, a place associated with enough impropriety for Mr. Darcy to be able to merely intimate: "*You* know him too well to doubt the rest".[29] In many of Austen's novels, as Elaine Jordan argues, the seaside is a place of chaos, where there is no class order, no propriety, no regularity.[30] *Mansfield Park*'s Fanny Price moves between the respectable world of the manor and the deprivation of Portsmouth. In *Persuasion*, Louisa Musgrove's excessive playfulness at Lyme Regis leads her to jump off the Cobb, rendering her unconscious and catalysing the novel's romances. Austen's *Sanditon* presents a striated social world that becomes dislodged on the sands. Andrew Davies's adaptation has the actor Theo James, playing Sidney Parker, emerge naked from the sea to the astonishment of the adventurous yet innocent Charlotte Heywood (played by Rose Williams). An escalation of Colin Firth-as-Darcy's infamous wet shirt scene, Parker's naked bathing is historically accurate: men did swim unclothed at the beginning of the nineteenth century and often not so far from family beaches. It is a scene seemingly too "modern" for Austen, yet just the kind of thing that this ironic and knowing author would have intimated as likely to occur on the notoriously unpredictable sandscape.

The association of sandscapes with play is tenacious, however scandalous this play might be. In the late eighteenth century, the philosopher and popular playwright Friedrich Schiller argued in his *On the Aesthetic Education of Man* that "man only plays when he is in the fullest sense of the word a human being, and he is only fully a human being when he plays".[31] The sandscape produces a playfulness that seems suitable to the writings and sketches in this book. Powys's *Weymouth Sands* reveals the sandscape as a place primarily of entertainment with its Punch and Judy show, the celebrated actor and the fortune teller. John Osborne's *The Entertainer* (1957) dramatises a tawdry mixture of sexual impropriety and the fading glory of the music hall in his character Archie Rice. Played by Laurence Olivier in Tony Richardson's 1960 film adaptation, Rice is an untalented womaniser who manipulates with an equal lack of success on stage and at home and reveals the fragility of pleasure.

Social danger and pleasure coincide often at the seaside, a mix that perhaps explains a backlash of inflexible morality. We see, for instance, the relentless boredom of the young Linda, a semi-autobiographical portrayal of Cynthia Payne, in a post-war English seaside town in the 1987 film *Wish You Were Here*.[32] The aridity of the town's rigid social mores is signalled in the film's opening shot of a grey, rainy sandscape on which a lone man walks his dog. In contrast, the bright face and billowing yellow dress of Linda signals vitality as she breezily cycles on the concrete promenade. Although she later suffers from the rigid imposition of respectability, such dreary legality proves to be ultimately ineffectual. The force of social codes separating good from bad becomes a farcical battle in *Carry on Girls* (1973), between a seaside town's moralistic mayor, the city council's resident feminist and the organiser of a beauty contest.[33]

It is often on the sands that moments of social disruption occur because social conventions can more easily be ignored in these shifting, shifty places. In the 1993 film, *Bhaji on the Beach* (Dir. Gurinder Chadha), it is on the sand that a normally reserved group of family and friends run into the sea and paddle. The older women, bastions of social propriety essential to the film's narrative, are seen shedding their normal customary behaviour. Even as individuals remain respectfully clothed in saris and long sand-coloured coats, they kick the surf gleefully. It is on the sands that shocking family revelations and moments of crisis takes place: an estranged father has a troubled conversation with his scared son and later drops his mask of gallantry when confronted by his wife before an audience of their extended families and friends. The sands are a place in which boundaries are challenged.

Granular Writing

This book attempts in its structure and style to reflect the granular and unpredictable nature of sand as well as its superlative creativity. After all, sandscapes are where we play. Adaptable and malleable, sand provides a place in which children delight in digging holes and building castles. It is fitting that, although this *Sandscapes* collection explores the cultural and historical, it also embraces different kinds of writing, from the noirish mystery of Angela Piccini's chapter, to the creative non-fiction of memoir and history in chapters by Jenn Ashworth, Brian Baker and Jean Sprackland. The recent boom in "new nature writing" has been marked by its approachable, imaginative style. Such creative writing is not only more accessible but enables us to imagine the life of things beyond human sight and understanding. Writing creatively allows us to better recognise the place of the imaginary in our everyday entanglements with the world. It is the space in which this felt, visceral, bodily, emotional and remembered engagement with the physical world might best be sketched out and reflected upon.

Poetics—a term that has its source in the Latin word for "making", *poesis*—has proved central to those who are writing on their relationship to this world of matter. Poetic language responds to the world in a different way from scientific knowledge. "Forces are manifest in poems", argues Gaston Bachelard, "that do not pass through the circuits of knowledge".[34] Poems allow access to an immediate, an unconscious and vibrant, response to the world. Poetry also allows us an imaginative leap outside of ourselves. Elizabeth Bishop's poem "The Sandpiper" offers us a bird's eye view of the sandscape that marginalises the human and draws attention to the autonomous energies of the world around us. We are introduced to the bird via the sound of the "roaring" of the sea that he merely takes for granted. The sandpiper is "finical" (the word from which we get "finickity", this insult to those who are interested in details). Bishop offers us a minute understanding of the bird's experience within the sandscape. The beach "hisses like fat", a domestic image that reproduces for human understanding this alien experience of the high-pitched tone of the movement of grains.

The creative has been the hallmark of even the most scientifically inflected nature writing. Karen Barad writes on the natural world and human-world relations through reference to quantum physics, that science that looks at life at the atomic level in which the physical world becomes strangely unpredictable. Barad argues that those who write on the "world's aliveness" must allow themselves to be "lured by curiosity, surprise, and wonder".[35] In a physical world that often defies expectations, creativity and creative writing are an important method for researching the world around us, especially apt for the peculiarities of the sandscape. Imagination is not only apparent in the literary, pictorial, musical or touristic representations of sandscapes, but aids our immediate encounters with them. How we see, hear, feel, touch, smell and—when it gets in our sandwiches—taste sand may only be appreciated if we attend to the textures of an encounter to which only imaginative and creative writing can, tentatively and albeit partially, get close.

Some of the essays collected here focus on the seaside town of Morecambe: Christopher Donaldson's chapter on travel across the Morecambe Bay sands to the Lakes; Jo Carruthers' chapter on the relation between class, sand, and seaside architecture; Shona Legaspi's fictional imagining of the life of a Bradford widow relocated as a B&B landlady; Angela Piccini's melding of Friuli, Northern Italy, and Morecambe in her murder mystery. On England's north-west coast, Morecambe is in many ways representative of the issues explored here.[36] Brought to national attention through the Chinese cockling disaster, it also draws visitors from across Britain because of the old-time glamorous appeal of the modernist Midland Hotel. Morecambe holds the majestic position of being #3 in the book *Crap Towns*, although it was once thought of as an upmarket gateway to the Lakes.[37] In the mid-nineteenth century, it became known as "Bradford-by-the-sea" because the new Bradford-Morecambe railway line, opened in 1850, catalysed the relocation of workers from Yorkshire during "wakes week". Full—even too full—of life in the summer months, it was quiet in the winter, as imagined in the chapter focusing on women's migration from Bradford to Morecambe by Shona Legaspi. Morecambe acts as a focal point, but the essays necessarily expand out to consider: the wider Morecambe Bay to Tresaith and imagined sandscapes (Jenn Ashworth); the neighbouring shore of Southport and Herman Melville's and Nathaniel Hawthorne's sojourn there (Jean Sprackland); as well as Elizabeth Gaskell's imagined Southport (Jo Carruthers); Essex sands/sounds (Brian Baker); Hampshire's Hayling Island (David Cooper and Michelle Green); a circular tour of Britain's shores (Sefryn Penrose); travel guides and the nation's sandscapes (Tim Cole); the imagined sandscapes of Forster's and Lawrence's novels (Nour Dakkak); an eclectic and ecological A–Z of sand (Peter Coates); and a global ecological agenda in the volume's final chapter (Julian Brigstocke).

Elements of sand's meanings weave their way through these chapters. The different authors reveal the peculiar, contradictory yet recurrent qualities of the sandscape: its secrecy and surprising revelations; the force of its elements and subjection to human force; an image of liminality and mobility; of instability and eternity. The collection begins with the novelist Jenn Ashworth's fragmented essay, which is part memoir, part reflection on the debris found on shore—tide wrack—and other matters "out of place" from sounds to selves. Ashworth's memoir provides a psychological resonance for found objects and waste that have received attention for their agency in Bennett's list of debris scoured from the sand at the opening of *Vibrant Matter*, and as revelations in Sprackland's chapter-by-chapter celebration of buried and washed-up objects, including tobacco, footprints and pollutants in her book *Strands*.[38]

Peter Coates gathers a written equivalent of the arenophile's more literal collection in his eclectic alphabet of sands. Coates offers a granular exploration of beaches, sand collectors and collections, with a focus on "gathering, removing, collecting, playing, treasuring, replenishing and saving" (p. 46). Sefryn Penrose's chapter has as its spine the transcript of a conversation with her mother about childhood beach holidays, from which she offers reflections and connections drawn from family diaries, etymologies, poetry and her own memories. The chapter by Tim Cole explores the mid-twentieth century as the era of automobile travel, comparing the poetic evaluation of sandscapes in mid-century Festival of Britain guidebooks and the shift to the coastal landscape favoured by the Royal Geographical Society. Cole traces the ways in which sandscapes became valued less as tourist sites and more for their unspoiled natural scenery; a shift from "beaches to be played on" to "cliff-tops to be walked upon" (p. 86). Nour Dakkak's chapter on queer sand explores how two Edwardian writers, Forster and Lawrence, align sandscapes with the ambiguity and unsettledness of male sexual identity. Portrayed as unsettled and shifting, sands allow these writers to reflect on the mutability of passion and sexuality. Brian Baker's essay on Essex sound/sand is set against soundtracks from The Who, The Triffids and the Cocteau Twins. Moments of silence, self-reflection and self-revelation intersect with the scenes of mod music, childhood Saturdays and family memoir. The meditative associations of the sandscape are invoked in Jo Carruthers' chapter, which compares elite, escapist experiences of the sands—as epitomised in the sleek Midland Hotel—and the tenacity of the seaside town's "rough" reputation. Reading the "rough" with the "smooth", Carruthers explores the part seaside style plays in social segregation. The Midland Hotel also features as the sandscape setting for Angela Piccini's "film noir", bringing us closer to its iconic interiors and Morecambe's imagined underworld. The story explores issues of labour and gender exploitation as it weaves together the northern shores of Morecambe and Friuli, Italy, and the historical tales of witchcraft that connect the two sandscapes. Shona Legaspi draws on oral testimonies collected in Morecambe for her fictional recreation of the experiences of Annie, a mid-nineteenth-century B&B landlady. Drawing on histories of women landladies, this retelling imagines the forces, influences and inspirations of sandscapes for those women such as Annie who made the journey west. Morecambe sands and the nineteenth century are also the location for Christopher Donaldson's chapter on travel "over sands" to the Lake District. Donaldson's essay draws on stories and accounts by different writers, namely William Wordsworth, Ann Radcliffe, Gaskell and Edwin Waugh, to reflect on the kinds of movement experienced on and through the Morecambe Bay sandscape—by foot, in cart, by rail. Moving further down on the map, Jean Sprackland returns to the sand dunes of Southport beach, the geographical focus of her book, *Strands*. Here, Sprackland focuses on the meeting of American writers Herman Melville and Nathaniel Hawthorne in 1856. Her pilgrimage to their meeting point on the shifting sands may be a "fool's errand", but by drawing on Melville's journal and Hawthorne's English notebooks, Sprackland reimagines the meeting that also provokes reflection on the exposure and shelter offered by the Southport sand dunes.

David Cooper and Michelle Green write about Hayling Island, near Portsmouth and consider the "diminution" of the sandscape, using the term diminution to denote retraction and a musical reduction-in-repetition. Identified as a "key reference point to the Anthropocenic imagination", the island reveals humanity's impact on the environment as part of the island that is only visible, briefly, at low tide and submerges further year after year. Cooper and

Green's essay provides a set of fragments—reflections, citations, documents, maps, memories and imaginings—for these seemingly unmappable and often overlooked sands. The impact of human presence on the world's sandscapes is contemplated in Julian Brigstocke's chapter which explores the relation between sand and power in the Anthropocene. Brigstocke's essay promotes a "thinking with sand" to find a "planetary ethic that welcomes the body of the earth into our experience of self and responsibility" (p. 210). Moving from the intricate qualities of sand to its global processes, this essay discusses drifting, saltation, fracking, erosion and mining as well as a planned Brexit lorry park in Dover.

This book does not expect to fully grasp or represent sand or sandscapes. It instead explores some of the multitude of aesthetic, historical, cultural and current political meanings of sand and sandscapes in studies ranging from sand dunes to sand mining, from seaside stories to shoreline architecture, from sand grains to global sand movements, from narratives of the setting up of B&Bs to stories of seaside decline. The essays attempt not only to see "a world in a grain of sand" but to think about how sand itself impresses itself upon the texture—or grain—of the world. Sandscapes are places of the imagination and here we attempt to imagine them anew.

Notes

1. Richard Cork, "Making Waves", *New Statesman* (August 2005), 28–29.
2. Studies tend to focus on the beach or on sand as a substance, rather than as terrain, as in this book. Notable works include: Ursula Kluwick and Virginia Richter, *The Beach in Anglophone Literatures and Cultures: Reading Littoral Spaces* (London: Routledge, 2015), a primarily visual arts and literary-critical study; Steven Braggs and Diane Harris, *Sun, Sea and Sand: The Great British Seaside Holiday* (Stroud: Tempus, 2006); Vince Beiser, *The World in a Grain: The Story of Sand and How It Transformed a Civilization* (London: Riverhead, 2019). Katie Ritson's beautifully written book is surprisingly inattentive to sand itself, focusing on rising sea levels and coastal erosion from a literary, cultural and ecological perspective that complements our approach here, see *The Shifting Sands of the North Sea Lowlands: Literary and Historical Imaginaries* (London: Routledge, 2018).
3. Nick Broomfield (dir.), *Ghosts* (Beverley Hills, CA: MySpace, 2006).
4. For histories of the early seaside, see H. G. Stokes, *The Very First History of the English Seaside* (London: Sylvan Press 1947), a journalistic and colourful study; and John K. Walton, *The English Seaside Resort: A Social History 1750–1914* (Leicester: Leicester University Press, 1983).

5. Walton, *The English Seaside Resort*, 22. See also Walton, *The British Seaside: Holidays and Resorts in the Twentieth Century* (Manchester: Manchester University Press, 2000).
6. Lewis Carroll, *Alice's Adventure in Wonderland and Through the Looking-Glass*, ed. Peter Hunt (New York: Oxford University Press, 2009), 20–21.
7. On the sea's health properties and Austen, see Jane Darcy, "Jane Austen's *Sanditon*, Doctors and the Rise of Sea-Bathing", *Persuasions On-line*, 38, no. 2 (Spring 2018).
8. Stokes, *The Very First History of the English Seaside*, 17.
9. Michael Bracewell and Linder, *I Know Where I'm Going: A Guide to Morecambe and Heysham* (London: Book Works, 2003), 56.
10. Daniel Burdsey, *Race, Place and the Seaside: Postcards from the Edge* (Basingstoke: Palgrave Macmillan, 2016), 19.
11. David Crouch, "Places Around Us: Embodied Lay Geographies in Leisure and Tourism", *Leisure Studies*, 19, no. 2 (January 2000): 63–76 (64).
12. F. J. Pettijohn, Paul Edwin Potter, and Raymond Siever, *Sand and Sandstone* (New York: Springer Science, 1987), 1.
13. See Beiser, *The World in a Grain*.
14. Steven Connor, "The Dust That Measures All Our Time", *Steven Connor* (May 2010), http://stevenconnor.com/sand/.
15. Jane Bennett, *Vibrant Matter: A Political Ecology of Things* (Durham, NC: Duke University Press, 2010), 13.
16. Nancy Scott, *The Seashore and Seashore Life*, illustrations by Jill Payne (London: Ladybird, 1964).
17. William Blake, "Auguries of Innocence", in *Selected Poems*, ed. Nicholas Shrimpton (Oxford: Oxford University Press, 2019), 77.
18. Kahlil Gibran, *Sea and Foam* (Milton Keynes: White Crow Books, 2009), 9.
19. Gordon S. Haight (ed.), *The George Eliot Letters* (New Haven: Yale University Press, 1954).
20. Jenn Ashworth, *Fell* (London: Sceptre, 2016), 113–114.
21. Matthew Arnold, "Dover Beach", in *New Poems* (Boston: Ticknor and Fields, 1867), 95–97.
22. Andrew Michael Hurley, *The Loney* (London: Tartarus Press, 2014), 323.
23. Wilkie Collins, *The Moonstone*, ed. Sandra Kemp (London: Penguin, 1998 [1868]), Ch. 4.
24. Sarah Perry, *The Essex Serpent* (London: Serpent's Tail, 2016).
25. E. Nesbit, *Five Children and It* (London: Puffin Books, 2008), 6.
26. See Lara Feigel and Alexandra Harris (eds.), *Modernism on Sea: Art and Culture at the British Seaside* (Witney: Peter Lang, 2009).
27. Ian McEwan, *On Chesil Beach* (London: Jonathan Cape, 2007).
28. John Bayley, "Life in the Head", *New York Review of Books* (28 March 1985).
29. Jane Austen, *Pride and Prejudice*, ed. James Kinsley (New York: Oxford University Press, 2004), 209.

30. Elaine Jordan, "Jane Austen Goes to the Seaside: *Sanditon*, English Identity and the 'West Indian' Schoolgirl", in *The Postcolonial Jane Austen*, ed. You-Me Park and Rajeswari Sunder Rajan (London: Routledge, 2000), 29–57.
31. Friedrich Schiller, *On The Aesthetic Education of Man in a Series of Letters* (Oxford: Clarendon Press, 1967 [1798]), 15.9.
32. David Leland (dir.), *Wish You Were Here* (London: Channel Four Films, 1987).
33. Gerald Thomas (dir.), *Carry on Girls* (London: The Rank Organisation, 1973).
34. Gaston Bachelard, *The Poetics of Space*, trans. Maria Jolas (Boston: Beacon Press, 1994), xxi.
35. Karen Barad, "On Touching—The Inhuman That Therefore I Am", *differences*, 23, no. 3 (December 2012): 206–223 (216).
36. Morecambe has been the focus of a number of seaside studies, including the book by Bracewell and Linder; also Karen Lloyd, *The Gathering Tide: A Journey Around the Edgelands of Morecambe Bay* (Glasgow: Saraband, 2016).
37. Dan Kieran and Sam Jordison, *Crap Towns: The 50 Worst Places to Live in the UK* (London: Boxtree, 2003).
38. See Bennett, "Thing-Power I: Debris", in *Vibrant Matter*, 29–31; Jean Sprackland, *Strands: A Year of Discoveries on the Beach* (London: Jonathan Cape, 2012). Increased attention to the lives and afterlives of found objects include the journal, *Discard Studies* at discardstudies.com.

2

Tide Wrack and Sand

Jenn Ashworth

Today I made myself at home in the tide wrack—that washed up rib of natural and manmade marine debris that marks the high water point on a beach. I couldn't write: the plot of the story I had been working on had become improbable and my interest in the characters had evaporated. Instead, I drove to Morecambe without a plan and ended up spending the precious work time between lunch and the afternoon school run picking through what beaches are made of. In the wrack and sand, I dug about with the toe of my trainer for something—inspiration?—and found only a broken bird skull, two waterlogged sanitary towels, a few faded beer cans half-filled with wet sand, a plastic cigarette lighter, some cockleshells and a sharp piece of terracotta-coloured plastic that looked like it came from a flower pot. Rubbish, that's what it was. Some mudlark I am.

A ragpicker—or in Baudelaire's Paris, a *chiffonier*—collects, grades, sorts and recycles the jumbled waste of the city. Before wheelie bins and public refuse collections he is the invisible street cleaner; the first iteration of the rag and bone man; the only one who picks up and takes away what remains at the end of the day. Baudelaire, in his poem "Le Vin de Chiffonniers" ("The Rag Picker's Wine") compared the figure of the ragpicker, lost in thought as he performs his dirty, lowly work, to a poet. In doing this, his work is dignified, sentimentalised and almost made sacred.

J. Ashworth (✉)
Lancaster University, Lancaster, UK

J. Carruthers and N. Dakkak (eds.), *Sandscapes*,
https://doi.org/10.1007/978-3-030-44780-9_2

As I wandered, sticking more or less to the high tide mark because that's where the interesting things are, I kicked through a heap of sweating seaweed knotted together with thread from an unravelling ribbon attached to a child's sock. There were two types of seaweed here: witchy bladderwrack (named after the dimples of the air chambers built into its fronds) and another one I don't have a word for. It had brown, limp leather hands, its dull surface strangely gelatinous. Seaweed inhales and filters everything: compounds, sometimes toxic, accumulate in its locks and silken or slimy stems. I was between Heysham and Sellafield: these plants are not for the plate.

It's more interesting to wonder not why the ragpicker is like a poet, but why the poet is like a rubbish collector, which is what Walter Benjamin did. For Benjamin, the method and the meaning of the poet/ragpicker's activities were bound together. And what a metaphor for writing this kind of rubbish collecting is: the curating, the filtering, the sifting and the sorting. The transformation of waste into art. The way it can be a living, sometimes, though not much of one. For a real poet, there's no such thing as rubbish.

Closer to the water, the sand cleared and became wet and colloid. It was fashionable today: *Apartment Therapy* Greige, the colour of filing cabinets. It tasted like fat and ash and salt and I dug for cockles but turned up only a ring pull. I reached the Midland Hotel which is weirdly still and pale like a clean white boat about to set off, or a spaceship, just landed. I've been in before, to eat little sandwiches and see the reliefs by child molester and artist Eric Gill but I'm not dressed for afternoon tea today. Instead, I headed into the West End and spent my last half hour haunting the charity shops, picking through loot from house clearances for a present for my daughter. I came away empty handed.

Out in the Black Rock Desert, where the sand is never wet like it is in Morecambe, the organisers of The Burning Man festival have a word or two to say about rubbish and wrack. To save wasting time on definitions or perhaps to allow no slipperiness of interpretation, they've developed a catch-all phrase for the remainder, the left behinds, the *stuff*—for all that might mar the perfection of sand after the festival, when the temporary Black Rock City is packed away and returns to whence it came: Matter Out Of Place (M.O.O.P.). M.O.O.P. is anything from a tent peg to a banana skin to a feather to the dark smudged ghost of a campfire on the playa (the word means "dry lake bed").

The unachievable ideal is for there to be no M.O.O.P. at all and the playa returned precisely to the state it was in before, "no physical trace of our activities wherever we gather"—as the website says. The festival springs up in the desert—a pop-up city, where the poet/ragpicker's final task is an act of both good citizenship and self-erasure. There on the playa—the flattest place on earth—no foreign object, not even wood or water, is benign. The Burning Man website warns that even a forgotten pencil left behind in the desert (*especially* a forgotten pencil?) can create a tiny obstacle for the wind-drifting sand to accrete against. It can pile up. It might, eventually, create a dune. For The Burning Man Festival's community, the work of the rubbish collector becomes a civic and moral duty. It is also a Sisyphean task, this, to make one of the largest and most famous arts and culture festivals in the entire world invisible: erasing the traces. But I'm struck by one of the most defining features of sand: the way it has a knack or a talent for being always in the wrong place itself—just ask the oyster. Its intrusiveness is intimate—we find it between our toes, painfully caught between folds of flesh after swimming, sticking to our fish and chippy tea on the promenade, discovered too late, as it grinds between our molars.

My mother had our carpets professionally cleaned, once, and the man who did them tipped out the water from his machine down the drain outside the house in the back yard and in the bottom of the bucket there was sand: grey Blackpool and Morecambe sand, probably—the exact colour of the cardboard backs of spiral-bound notebooks. The soup slopped down the drain and I thought of all our beach trips, the rubbishy remnants of all our sad days out caught in our boot-treads and carried into the house, hanging out in the carpet fibres until the man, swilling his bucket under the outside tap, accused my mother, only half-joking, of having shares in Shake-n-Vac. Either now or then or both I remember the advert for it, and its jingle, where a desperate woman in a white skirt prances and gyrates around her living room with her vacuum cleaner, as she banished embarrassing odours from her carpet. She uses Shake-n-Vac *to put the freshness back*, and this song is an earworm, another kind of persistent Matter Out Of Place: the aural and psychic equivalent of a stone in a shoe or a piece of grit between lid and eyeball. And after I return home to Lancaster, I will learn again about sand's persistence: it becomes Matter Out Of Place in my shoes, in the footwell of the car, even after valeting, and under a fingernail forever and ever and ever. The beach trip taken in working hours should have been a secret, but my children worked it out immediately.

There's no pearl pure and pretty enough to pay for the damage a grain of sand, held in the head since childhood, can do to the mind. It is deceptive though, the way sand can feel as if it's nothing but the perfection of its surface: the regularity of it, the Zen smoothness of its tide-raked ripples. What lies beneath? What does that sandy perfection hide? Ask the Morecambe Bay cockle pickers, dead in the water, and all those that stuck and panicked and died before them, discovering the suck and pull of tidelands and saltmarsh. Not for nothing did Tim Burton in his film *Beetlejuice* trap the dead Maitlands at home for a century by making the entire world beyond their own front door into hell, a hell without fire, but furnished only with yellow sand that stuck to their clothes and provided a home for the sand-worms that surfed through the dry landscape, coiling and uncoiling, biting, predatory, waiting. (Writing is like this: the entire world shrinks to your own house, and there are monsters at the door.)

If I can't work out the story, I should instead do a research paper about things that live in sand. I could use it as an alibi to do what I have been doing anyway, which is spending a period of compassionate leave from work watching the entire *Tremors* franchise of films. In the world of *Tremors*, sand-worms are called graboids and they tunnel through the inland desert of the sandy Nevada landscape. Perfection, Nevada, is a little settlement (hardly a town—in the first film, the good one, there's only 14 inhabitants of the place first named Rejection) tucked into a box canyon.

I'm familiar with this landscape: grew up seeing animated cartoon recon-structions of the Mormon trek west, along emigration trails that cut through the sandy badlands of the American south-west. I did not possess that land-scape, but it possessed me and in that way it was a home from home. A kind of Mormon Mecca to which the good girls who married American mission-aries would be spirited away someday. There's a long and complex history to this which meant that the first time I visited Salt Lake city, on being asked where I came from, I said, "Preston," and was about to explain where it was when the Mormons behind the counter at the bookstore nodded and smiled with homely enthusiasm. *Of course* they knew where Preston was. The very first missionaries landed on the North West coast and the first English convert girls had set off from the docks at Liverpool—polygamous commu-nities need lots of wives, of course, though they weren't told this until the land had disappeared from view behind them. Never mind. The old maids that stayed behind, waving goodbye from the shore, built a satellite of Zion on their own North West coast: there are more Mormons than you'd think in Preston, Southport, Blackpool, Lancaster and Morecambe. Life's a beach.

Anyway. *Tremors*, which is what I have been watching because I have been too sad to write, is a bromance and the two male leads—Val McKee and Earl Bassett—are handymen on the make. A right pair of *chiffoniers* they are, loading up their flatbed truck with scrap, enjoying mishaps with septic tanks, and taking some kind of pride in the dirty work. "Nobody handles garbage better than we do," says Earl, taking a chilled beer from a toilet bowl repurposed as an ice bucket. When the sandy earth starts to give up its monsters, these two are the obvious choice to deal with the problem.

A short diversion while we think about these monsters, their methods and what they tell us about sand. For while it is true that the "graboids" chew and spit—their seeping, mucusy, many-toothed mouths are an object lesson in abjection (vagina dentata, anyone? They're repeatedly referred to as "one big mother"—but *of course* they are. And, yes, it is worth reminding ourselves that even the word "abject" contains an idea about rubbish or litter, coming as it does from the Latin word *abicere* which in English means "to throw away")—their main way of killing is to grab the foot of a victim who walks upon the sand and pull him underground. Literally, to be caught by the graboid is to have the ground swallow you up. Cars too—whole cars sometimes, for the thin and loose Nevada soil is capacious.

To be swallowed up—or to wish for it, to wish for the ground to open and swallow you up—to be caught by a graboid—is something to do with embarrassment, isn't it? Embarrassment—a kind of misstep or gentle sinning (for your embarrassing act or state doesn't contain the wrongness, only the context in which you place it), partly to do with inappropriateness, being in the wrong place, or the wrong time. All of the inhabitants of Perfection/Rejection are embarrassed in some way: they're poor and hopeless, the mining industry is gone and has left them behind, they are crazy misfit survivalists and most of all, they are litter, or matter out of place. They should not be there.

Tremors consistently and constantly remarks upon the inhospitability of the Nevada desert (this inland beach). It is partly because the plot relies on geographical isolation and the entrapping aspect of the box canyon's topography, yes—but these constant reminders (the sad sprinkler attempting to wet the parched ground in front of Nancy Sterngood's trailer, Old Fred's sheep corralled in a sandy field with no grazing to be had for miles, the numerous monster-eye shots of the ground as it is rapidly tunnelled through: the sand and grit tumbling in the darkness, reminiscent of the way the stars seem to part and flow past the viewer during the opening of the Star Wars films—there's a link here between sand and rubbish too—Rey, the scavenger of a desert planet—and of course if only there were time and life enough for *Dune*) of inhospitability say something else: the graboids aren't the foreign bodies here, the people are. This sand is not a place for humans. It might look like the ground is swallowing them up, but this is only the relief that happens after noticing you're in the wrong place.

Later on, insomniac, and still unable to write, I iMessage my Australian friend who is drinking her morning coffee and just about to start her sunny day. I tell her about my beach trip. About the fact that it didn't help. I tell her that when I was little and we went on days out to Blackpool I'd dig holes on the beach to bury my father, or to watch the frothy water rise up like magic from the bottom, or in productive moods, to shovel my way through to Australia. I had no concept of the country, other than everything was upside down there and, like Narnia, the Australian winter was uninterrupted by Christmas. She's home for the winter and she promises to send me a picture of Santa on the beach in his board shorts to show my children, who do not believe that the sun shines on Christmas day. I ask her where Australian children think they are digging to when they make holes on the beach, and she laughs at me and tells me that they don't. The digging might be, we theorise, a peculiarly British occupation.

There's a piece of biomythography (I pluck the word from Audre Lorde's hand) circulating in my family about both Australia and Blackpool. One of those near miss, might-have-been type of stories. I don't even know how true it is, but apparently, if it weren't for the pull of the 1960s bingo halls and the disreputable delights of the sea-front arcades my Glaswegian grandparents enjoyed during the annual Glasgow Fair—the Scottish equivalent of the Wakes weeks, an annual synchronised holiday for industrial communities in the north (for a while, Morecambe was known as Bradford-on-Sea, because all the millworkers would decant there each August, whole streets of families reconstructing their neighbourly relations in rows of sea-front B + Bs), they'd have gone "Down Under" as some of the last ten pound poms. Which means, in some world through the looking glass, I'd have been Australian and not my father's daughter. But my Granny loved Blackpool so much they moved to Preston, where my Granddad started working in a paper factory and they lived their first English years in a little council house on a street named after Dunbar, the Scottish poet.

Sometime in 2005. It's raining, then it stops, then it starts again. A day like that. A week like that. Whole months like that. And bored of waiting for fine weather, we've come out to Squires Gate to walk. I haven't slept properly in weeks—no, months—but I allow myself to be dragged out because the fresh sea air will cure what ails me. The wheels of the expensive all-terrain pram get caught in the sand almost immediately and I have to carry the baby while the man who I am with—we are not really together, and we won't even try to be for much longer—folds the pram up and humps it back, over the dunes. I wait. The baby is wearing a pale green snowsuit with a hood. The bundle I carry doesn't feel like a baby, it feels like a package of clothes, one of those large padded jiffy-bags stuffed with impulse buys of clothing from ebay. It is too soft and too heavy.

From between the dunes, a man emerges. He has a dog with him, and bends towards it, to release its collar from the lead. The wind is blowing my hair into my eyes and mouth and because I am holding the baby I cannot move it away but I can still see the dog squat in the sand and shit. The man looks at the dog. Looks around. Walks away. I'm walking now. Running, with the baby jogging in my arms, its head against my shoulder. I shout. I am not supporting the baby's head properly. The man who I am with will tell me that later, once he's threaded his way back through the dunes. He'll say I was running, and not supporting the baby's head properly. He says it now—*you're not supporting her head properly!* but I run away from him and away from the man with the shitting dog and sit in the damp foothills of the dunes. I put the baby in the cold sand next to me. It's crying. Both of us are.

The man I am with picks up his daughter and brushes the sand from the back of her snowsuit. *He shouldn't have let the dog shit in the sand*, I keep saying. I'm thinking of Tommy from *Trainspotting*, dead in his flat of Toxoplasmosis. That's cat shit though, isn't it? It seems very important to find out as soon as I can but he says *come on now, come on it doesn't matter, the baby's cold now let's just go home we can try again another day* and so we turn and climb the dunes, our feet sinking into the drifts of cold sand. Only the top layer is wet, and the underneath is dry and fragile and we're breaking the dunes by climbing them, but it's the shortest way, and so we do. He has a tumour in his gut, this man. Right now. He doesn't know. Nobody knows. But it is there.

The rain comes on harder and I want to run but I can't because running up the side of a dune feels the way running does in a nightmare, when the monster is coming, and your feet are sticking to or sinking into the ground (there's a bit in *Nightmare on Elm Street* about this, I think, where Nancy is trying to run up the stairs to get away from Freddy, and the stairs become colloid too, like wet sand, like this, and she sticks to them and sinks into them and has time to look at the ooze between the sole of her foot and the collapsing stair tread as she despairingly climbs upwards, and there's no point trying to find the YouTube clip because everybody knows what a bad dream feels like—what it feels like to try and wake from one, and be unable to) and the sand is pouring into my trainers and it is damp between my feet and my socks, and now it is inside my socks, cold, and scraping at my skin, and it hurts and it's horrible and makes me cry more. I can't stop crying so I should stop walking and turn back, but this is the quickest way from the beach to the train station, so we keep on, the wind throwing the sand into our faces and the baby's cry throbbing in all the soft parts of me: my throat, my eyeballs, my breasts.

Matter Out Of Place isn't a phrase the organisers of the Burning Man festival invented, of course. It comes from Mary Douglas's book *Purity and Danger* and was a way of thinking about foreign objects, about cleanliness and contamination. About the way our ideas of what is wholesome and what is taboo are only ways of marking out boundaries between the things we want and the things we don't want—or rather, the things that are part of us and the things that are not part of us. Or rather, the things we don't mind being part of us and the things we do mind being part of us.

These are impossible boundaries to maintain, of course. Wishes, not facts. Consider cancer, when it becomes so entwined with the inner landscapes of the body that it is impossible to remove. Consider how we decide who is a member of our family and who is not.

2017. We've taken the children to Tresaith beach. We're in grief, all of us, but my daughter in particular. We satellite around her, trying to be present and ready, but not to crowd or irritate. She lies on a towel scooping and smoothing the sand. We give our children tools and encourage them to burrow and sculpt in drifts of broken stone and the crumbled exoskeletons of dead invertebrates. And they do.

"We took Daddy with us when we went to Devon," she says. She doesn't look at me: she's staring at the sand between her fingers. She smooths out an area in front of her towel into a writing surface, draws a pattern into it with her index finger, then flattens it out again.

The summer has been a rotating series of trips to the beach with immediate and wider family. We're all trying to give her a treat. To distract her. She's suddenly too old for sandcastles and none of us know if this is grief or being twelve.

"Did you?" I say, carefully.

"I thought it would be fancier. The jar. It was red plastic."

"Did it have his name written on it?" I don't know why I ask this.

"On a sticker", she says, with such disgust and contempt that I want to laugh.

"What was it like, taking him with you to Devon?"

Our conversations are often like this. She offers me an observation or a fact—like a ball tossed—and I either have to hold it in my hands—warily (sometimes these balls turn into grenades)—or try to find a way to hand it back to her.

"It was weird," she says. She doesn't look at me, but carries on with the sand, writing and flattening out, writing and flattening out. I want to change position so I can see what she's writing but if I do she will stop, or get up and walk away all together. Conversation with my daughter this summer is like stalking a sparrow.

"Weird."

"Weird?" (It is a therapist's trick, this. It fucks me off when people do it to me, and I do it to her. Weird. Weird?) I wait.

"Inside. We took the lid off and inside it was like shredded newspaper and broken up bits of stone."

She writes. Erases. Writes again.

"I didn't think it would be like that. It would have been good to see a picture of what it looks like so people who take the lid off aren't surprised."

I try to imagine this. Shredded newspaper. To visualise what she has seen. I haven't seen it. I'm not listening in the right way. It's my job to sift through what she tells me to find the kernel of the thing she wants me to know. She is telling me she was surprised. That it was a shock. That she keeps on being surprised. She stirs and sifts the sand. Buries one hand in it, right up to the wrist, then twitches her fingers. The surface cracks, falls apart. She pats it back into place again.

"I think I would have been surprised too", I say.

She gets up, sand sticking to her legs and arms, scattering over the towel and flying as she runs towards the sea.

One day somebody will write an entire book about sand. The cover, designed as cheaply as possible by an intern just getting to grips with Photoshop, will, against the express wishes of the author, use some version of that photograph that went around Facebook and Twitter a few years ago. The one that showed what individual gains of sand look like viewed at ×1000 magnification. How special and different and beautiful each grain is, etc. This book's main work will be to advance a fresh taxonomy for the grading of grains. It will describe new units of measurement invented solely for the task of determining minute gradations in size. It will outline the means of measuring and recording results and the margins of error for various types of instruments: mainly to do with the reliability of their magnification in lab and field conditions. The author won't say this in so many words, but there will be a strong suggestion that the practice is more of an art than a science. *There's a knack to it*, he'll say, during the sparsely attended speaking engagements arranged around the publication date of the book.

The book will also go on to describe various ways the shape of individual grains (their roundness or angularity) can be determined and described. People will comment on the uniqueness of this. Never before has such detail—gathered and collated from literature surveys and years of field-work—been provided on the materiality of particles, they'll say. There will be tables listing the eight types of surface texture and factors resulting from it, for example, skid resistance, and other pertinent properties. There will be pages about hybridity, and how to determine the cross-types of both individual grains and a larger quantity of sand. The book will be very expensive to produce because the author will insist on a great number of colour photographs, which will limit the publishers willing and able to consider such a project. The author will be happy enough with the publisher finally willing to take it on, though once or twice he will allow himself to feel the extent of his modest wound: of course it would have been nice if the manuscript had been accepted by one of the larger university presses, he'll think sometimes, at night.

The final chapter of the book, included against the advice of his editor, will undermine (in a move one major reviewer will describe as "misguided" and another as "poorly conceived") what has been written before by outlining (evidence will be provided to back up all points, provided in a list of numbered tables and diagrams and additional material found in a series of lettered appendixes that will further increase the production costs of the volume) the defining feature of sand and sandiness. Located not in the nature of the grains themselves, the author will conclude, but in the quality or properties of the space that lies between them. Found in the way the particles—suspended in air or water—map out a three-dimensional network. The way they bump and slide past each other, or the air that holds them, in constant contact, but only tangentially so. To some readers, it will appear that (between the lines) the author is confessing he's spent his entire career looking at the wrong thing.

The author will gift a signed first and only edition hardback to his mother, who will never read it. A year after publication, the remainder of the print run will be pulped in an out of town recycling plant near Warrington.

References

Burning Man. "Matter Out of Place (MOOP)." Accessed May 5, 2017. https://burningman.org/event/preparation/leaving-no-trace/moop/.
Burton, Tim (dir.), *Beetlejuice*. New York: The Geffen Company, 1998.
Craven, Wes (dir.), *Nightmare on Elm Street*. New York: New Line Cinema, 1984.
Douglass, Mary, *Purity and Danger: An Analysis of Concepts of Pollution and Taboo*. London: Routledge Classics, 2002.
Underwood, Ron (dir,), *Tremors*. Hollywood: Stampede Entertainment, 1990.

3

An Eclectic A–Z of Sand: Removing, Treasuring, Recreating and Protecting

Peter Coates

A is for Arenophile

Accidental sand collection—in shoes and clothes, or between our toes—is an unavoidable by-product of a beach visit. But it can also be purposeful. Arenophiles collect samples of sand, a substance varying enormously in texture, angularity and colour as well as mineralogy and locale (which include deserts, lake and riverbeds as well as seashores). Arenophile stems from *arena/harena*, Latin for sand—retained in Spanish but replaced by *sabulum* in French *sable* and Italian *sabbia*. (*A synonym, Psammophile, is the chosen entry for P.*) *Arena* then doubled as the word for stadium because sand was strewn across Roman coliseums and other sporting amphitheatres to soak up spilled gladiatorial (and animal) blood.[1]

Pursuing arenophilia requires minimal equipment: re-sealable plastic bags or film canisters, a spoon, a magnifying glass, a notebook and (back home) a display cabinet (a pill organiser works nicely). Collectors—whose hobby has spawned a multiplicity of websites and chatrooms[2]—range from sedimentologists to vacationers, the spectrum of ambition stretching from systematic sand gathering from every nation or desert to occasional beach holiday samples.[3]

To acquire rare samples (star sand, for instance, and rarer colours), arenophiles, like philatelists, trade with fellow enthusiasts. Green sand, for example, exists in only a handful of known places, including Papakolea,

P. Coates (✉)
University of Bristol, Bristol, UK

J. Carruthers and N. Dakkak (eds.), *Sandscapes*,
https://doi.org/10.1007/978-3-030-44780-9_3

Hawaii, and Hornindalsvatnet Lake, Norway. The business of collecting is regulated and supported by the International Sand Collectors Society, which advises members of legal restrictions, facilitates swaps and arranges an annual conference, SandFest.[4] The Society's motto: discovering the world, grain by grain.

B is for Biogenic Sand

Rock is not the parent substance of all sand. Some derive from biological rather than mineral (abiogenic) sources. These sands (aka organic, calcium or biological sands) are composed of the usually light-coloured remains of bone and other hard (calcareous) bits (rubble) of sea creatures such as barnacles, clams, corals, coralline algae, echinoids (sea urchins, whose purple, black and green spines add dots of colour), foraminifera, sea snails and sponge spicules. A lucky beachcomber may find a tiny tooth, jawbone fragment or even an entire sea urchin spine. Grains of biogenic sand come in all shapes and sizes. The skeletal remains of plants also contribute. Some sand is a mixture of mineral and biogenic grains: the majority component determines whether a beach qualifies as mineral or biogenic. Yet certain isolated tropical beaches are exclusively biogenic. The pinkness of Bermudan beaches derives from pink, single-celled foraminifera, but coral sand is an often misleading label for such beaches. For one-hundred per cent coral sands rarely exist and sands that are mostly coral are confined to beaches near coral reefs occurring in warm waters between 30 degrees N and 30 degrees S.

The proportion of biogenic materials tends to be higher on island beaches, and another biogenic source for Hawaiian beaches is fish excrement: parrot-fish eat coral and excrete the pellet-like waste. To establish whether a sand sample is biogenic or abiogenic, douse with a pinch of vinegar. If it contains bones or other skeletal remains, the acid will react with calcium carbonate to produce bubbles of carbon dioxide.

C is for Cleopatra's Beach

Cleopatra has three Needles (London, New York and Paris), but just one beach. Yet unlike the Egyptian obelisks re-erected far from their places of origin in the nineteenth century, this alluring beach, which every guide-book to the Turkish Aegean highlights, enjoys a genuine connection with the Ptolemaic dynasty's last ruler. The sands of Cleopatra's Beach, a tiny cove at

Gökova Bay on Sedir Island, consist of grains of creamy white, silky smooth, calcareous Holocene ooids, most of them perfectly spherical and the size of a grain of white sugar. These exotic sands are confined, though, to a surface layer no more than 80 centimetres thick at the beach's seaward end, vanishing completely a few metres beyond low-tide line. Nowhere else in the Aegean can you find this particular sand type.

According to legend (but unmentioned in any ancient text), Sedir Island was the secret meeting place for Roman ruler Mark Antony and the Egyptian queen. Circa 35-32 BC, to delight his lover, Antony indulged in an extravagant act of beach improvement by importing sand from the Egyptian coast west of Alexandria (Cleopatra's hometown and capital). Recent "provenance analysis" of sand grains' microfacies qualities (under a microscope) reveals a close match between Cleopatra's Beach and Al Agami beach, 12 kilometres from Alexandria, and part of the only stretch of Mediterranean shoreline with an ample quantity of this distinctive sand. Al Agami is also located more or less due south of Sedir Island: according to a sand specialist, fifteen fully loaded Roman corn barges would have been required to bring over the quantity of oolitic sand (180,000 tons) on Cleopatra's Beach in the late 1980s.[5] Sedimentological data also rule out the likelihood of in situ formation— protected Gökova Bay is a site of "low energy" wave and tidal activity, which also accounts for the Egyptian sand's survival. Research additionally reveals that these sands show "evidence of being out of place" on Cleopatra's Beach, and that the case for its origin as ship's ballast is unpersuasive.[6]

The sand comprising what the signs hail as this "unique beach" is so precious that a stone retaining wall encircles its landward sides. Sand removal is forbidden; and to further minimise loss, swimmers can only access the water via a wooden pathway parallel to and outside the wall; towels, to which sand might cling, are also prohibited.

Cleopatra's Beach could be the world's first artificial beach (human-made beaches now range from the sands of Barceloneta, part of Barcelona's redevelopment for the 1992 Summer Olympics, to the more recent urban beach phenomenon of riverside cities like Berlin, Paris and Newcastle). More importantly, that incongruous patch of sand in the Aegean reminds us of the strength of our desire to shape and re-shape our sandscapes and seashores— not to mention sand's possession of that infinite variety that Enobarbus attributed in Shakespeare's *Antony and Cleopatra* to her who inspired the relocation of sand across the Mediterranean.

D is for Dredging

The demand for the right kind of sand (see entry for R) brings massive excavations on land and underwater. Dubai's offshore creations, Palm Jumeirah and the World—the latter made up of 300 islands—are a stark example of how undersea sand is increasingly scooped up to create land.[7] Dredging the Gulf of Persia's seabed brings the "undersea equivalent of choking sandstorms" that wipe out marine organisms and coral reefs as well as transforming water circulation patterns.[8] Sand mining (a more appropriate word than dredging) is driven by rampant urban growth worldwide: sand, according to Vince Beiser, is "to cities what flour is to bread".[9] Associated with a lucrative illegal trade, mushrooming cities are wrecking sandy habitats worldwide, freshwater and marine, including China's largest freshwater lake. Lake Poyang's sand is barged down the Yangtze to make the concrete and asphalt of burgeoning Shanghai. The previous source of sand, the Yangtze's bed, was abandoned when bridges were undermined and sections of riverbank collapsed. Over 30 times more sand than rivers contribute annually to Poyang is currently being gouged out of the lakebed, ruining local fishermen's livelihoods.[10]

While sand is being appropriated to facilitate Dubai's expensive new real estate, elsewhere, like Indonesia, sandy islands are literally vanishing, sacrificed to Singapore's urge to increase its territorial area. In California, protestors fight the last of the sand mines that were a common feature of the state's coastal sandscapes until the 1980s, when federal legislation shut all of them down, except Marina's. The CEMEX Lapis plant (est. 1906) survived there thanks to a legal loophole—location in a lagoon above the mean high tide line placed it beyond federal jurisdiction.[11] In January 2017, at Marina state beach near Monterey, activists carrying placards reading "Take a stand, save our sand" emptied 200 pounds of bagged sand bought from a building supply store onto the depleted beach from which it had been removed. In Britain, dredging from the bed of Northern Ireland's Lough Neagh, an internationally important site for breeding and migratory birds and site of a valuable commercial fishery, also faces opposition. These activists are not arenophiles, conventionally defined, but value sand, in situ, as habitat for nonhuman creatures, the basis for local economies and to slow down the rate of coastal erosion.

E is for Extra-Terrestrial Sand

Timanfaya National Park in the Canaries, part of the Lanzarote and Chinijo Islands Geopark, consists of volcanic cones and bowls, variegated sands and lava fields stemming from an almost unbroken string of eruptions between 1730 and 1736. This unearthly landscape provides scientists from the European Space Agency with a research site unique within UNESCO's Global Geoparks Network: a so-called planetary analogue. Since Timanfaya is the closest terrestrial match for Mars, ESA's Pangaea programme, whose goal is to send European astronauts to Mars, has been training here since 2016. As Spanish geologist and astrobiologist Jesús Martínez-Frías observes, Lanzarote is "Mars on Earth".[12] One of the features that create this geological and environmental approximation to other planets is Lanzarote's multi-coloured sands. Mars, named for the Roman god of war, is commonly referred to as the Red Planet because of the iron oxide its rocks contain that lend its surface sands a rusty hue. Dunes are a common ingredient of the Martian sandscape (though their mobility is much more muted than their earthly counterparts' since wind energy is substantially weaker there than on earth).[13] The dune complex on Mars that has become most familiar to us earthlings, through images taken in 2014 by NASA's "Curiosity" mega rover, is the Dingo Gap sand dune, which rises to just a metre.

F is for Fennec Fox

This smallest of canids—the average domestic cat is bigger—reminds us that sand is not just on the beach. The habitat of the fennec fox—Algeria's national symbol and its soccer team's nickname—stretches from the western Sahara to Kuwait in the east, including Israel's Arava Desert. Specialised features allow it to thrive in its sandy environment, where it burrows dens in the sand of stable dunes—the more compacted the sand, the more extensive and complex the dens. Like the sand dune cat, super-furry feet protect its paw pads from being scorched. Enormous ears (bigger than those of the bat-eared fox, and another feature shared with the sand cat) allow it to detect prey (insects, rodents and reptiles) lurking under the sand, as well as providing a cooling mechanism. Thick fur keeps it warm during chilly desert nights and gives extra traction when moving over loose dunes. So finely attuned is the fennec fox to its sandy world, it might as well be called the sand fox.

G is for Goodwin Sands

Goodwin Sands, 3.5 miles off the coast of Kent, is a sandbank 12 miles long and 5 miles at its widest point, built up by tidal rise and fall within the narrow neck of water known as the Dover Straits. At high tide, they are totally submerged; at low tide, as much as a tenth of the overall area is exposed, with ridges up to 13 feet above sea level. A ship that runs aground either partially or fully is liable to break its back as the tide ebbs (as much as 18 feet in spring), leaving the bow or stern unsupported. The proximity of one of the world's busiest shipping lanes has earned these shifting sands notoriety as the graveyard of ships. Over 2000 wrecks have been recorded. During the Great Storm of 1703, four ships came to grief, with the loss of 1200 sailors in a single night. The peculiar tradition of playing cricket there at low tide in summer began in 1824 and survived until 2003, and equally bizarre proposals to locate a third London airport on the sands were floated in 1974. A recent decision to grant Dover Harbour Board a licence to dredge its subtidal sand prompted the formation of a campaign group. Goodwin Sands SOS wants to protect the site's rich underwater cultural heritage, its wildlife resources, which include hundreds of grey and harbour seals, as well as the Sands' increasingly vital role as a natural sea defence for the vulnerable east Kent coastline. In a world of rising sea levels, sandbanks can be a blessing as well as a bane.

H is for Hourglass

In ancient Babylon and Egypt, the water clock (clepsydra) told the time. We do not know precisely (or even roughly) when sand superseded water, but the earliest visual rendition is probably within Ambrogio Lorenzetti's fresco (Siena), *Allegory of Good Government* (c. 1338). This detail depicts Temperance holding a large hourglass with sand (aka sandglass) in her right palm (the hourglass's glass, like all glass, is melted sand). The British Museum's Horological Collection houses the earliest surviving hourglass (probably German, 1520).[14] The sand used for this purpose was clean, dry and carefully sieved to ensure that grains were of more or less similar size, and preferably smoothly spherical to ensure consistent flow; rate of time passage was determined by the quality and fineness of the sand. Often, though, it was hard to tell whether the so-called sand was actual sand (natural silica) or powdered marble, tin ore, lead oxides or even pulverised eggshell.[15]

Maritime travel in the fourteenth century was the first area of activity that widely adopted the hourglass (marine sandglass). Sand was preferred to water aboard ship because, as well as not freezing, its flow was less prone to disturbance by rocking and pitching.[16] Also, whereas water clocks were susceptible to the blurring of clarity from condensation resulting from temperature change, if tightly sealed, sand was unaffected. (Nowadays, the hourglass is most frequently encountered in a sauna.) Non-terrestrial deployment was quickly followed by use on terra firma to measure things such as the length of a sermon and cooking times. Though cheaper and easier to use than the first mechanical clocks, from c. 1500, the hourglass lost ground as the former became smaller, more accurate and more affordable. Despite its name, the hourglass could measure shorter periods of time (witness its continuing use as an egg-timer), and a long period "hour" glass can measure up to 24 hours.

The continuing fascination the hourglass exerts may have something to do with its capacity to provide a tangible visualisation of the past and the future, how the narrow waist of the present separates them, and how "the sands of time" eventually run out for us all.

I is for Iwo Jima

The five-week battle of Iwo Jima was fought on the beaches of this second largest of the Japanese Volcanic Islands between 19 February and 26 March 1945. The American invasion (Operation Detachment), whose objective was to capture the whole island, to provide a launch pad for assaults on the main Japanese islands to the north, triggered one of the bloodiest conflicts of the War in the Pacific, with 6800 US and nearly 18,000 Japanese deaths and thousands of casualties. Among the movies the battle has inspired are *Sands of Iwo Jima* (1949, starring John Wayne) and Clint Eastwood's *Letters from Iwo Jima* (2006), originally entitled, more evocatively, *Red Sun, Black Sand*.[17] American veterans returning to Iwo Jima's beaches developed the habit of pocketing mementoes of the black volcanic sand that hindered their progress. One veteran compared it to "walking through wet coffee grounds".[18] Ninety-year-old Frank Pontisso, a former Marine who returned in 2015 to mark the 70th anniversary, packed a 20-ounce plastic soft drinks bottle with the super-fine grains of black sand from places such as Green Beach, where he first landed (to distribute among friends back home in Des Moines, Iowa).[19] In Pennsylvania, the Gettysburg Museum of History's gift shop sells authenticated vials of Iwo Jima sand a Marine collected in the 1980s, and bags containing two teaspoons-worth (plus validation picture) are advertised on eBay.

While this sort of sand collecting may be a harmless activity, larger-scale removal constitutes theft. In August 2019, police discovered 40 kg of white sand from the beach at Chia, Sardinia, jammed into fourteen plastic bottles, in the trunk of a vehicle boarding the ferry from Porto Torres to Toulon, France. The couple could face a fine of up to €3000 and a prison term of up to six years under legislation (2017) outlawing removal of the sand that represents a vital ecological asset and tourist attraction.

J is for Jaywick Sands

In the early 1930s, a few miles from the well-established Essex resort of Clacton-on-Sea, and 60 miles northeast of London, a local developer sold small parcels of land on the North Sea coast ("a certain number of cheap grass plots adjoining the sea at Jaywick Sands Estate are offered….£25 freehold. A chalet can be erected for £20").[20] The buyers were working-class Londoners (hence the soubriquet "Stepney-by-the-sea"), not least those with jobs at the recently opened Ford factory at Dagenham. The village of glorified beach huts, chalets and bungalows that sprouted up on the saltmarshes and grazing lands of Jay Wick Farm became known as Jaywick Sands.[21] "Come and see Jaywick, the Model Seaside BUNGALOW TOWN, where you may own a freehold chalet right by the sea for LESS THAN THE COST OF A CARA-VAN", urged a 1934 advert.[22] Though ill-equipped for year-round residence (many dwellings lacked heating, running water and proper drainage), many holiday homes became permanent homes for those bombed out of their East End communities during the Blitz. Decaying buildings and unpaved, potholed roads were just the most visible signs of blight in the village's steady decline since the late 1940s. In 2010 and again in 2015, the Index of Multiple Deprivation (Department for Communities and Local Government) designated the eastern half of Jaywick Sands as the most deprived area in England; youth unemployment, for instance, is the nation's highest.[23] To make matters worse, the risk of flooding the village faces is the most acute in England.

The golden sanded, crescent-shaped beach that originally attracted Londoners mobilised by private automobile ownership—part of a seven-mile stretch of sandy coastline—rarely if ever features in Jaywick's heavy media coverage triggered by the 2015 report.[24] Early 1930s photographs show deckchair dotted sands and bathing costumed fitness enthusiasts stretching their limbs on the beach.[25] Wild swimming advocate Roger Deakin recalled family holidays in a rented wooden shack on stilts directly fronting the beach shortly after World War II ("it was like living in our own sandcastle").[26] Christine Lee remembered a family holiday around the same time at the chalet her father bought in 1939. She and her sister built "ever-larger sand-castles, and buried each other up to our necks in cool, damp sand".[27] The adjacent Martello Beach Holiday Park still provides caravan accommodation and pitches for camping and the sands remain golden. Despite inclusion on Deakin's wild swimming map of Britain, Jaywick Sands, like many other British seaside resorts, is unlikely to enjoy a reprise of its interwar and post-war golden age.

K is for Kahana Beach

Golden sanded Kahana Beach on Maui's west coast is one of Hawaii's best-known beaches. Yet it is also one of Hawaii's most threatened, having suffered drastic sand loss through a combination of rising sea level, increasingly frequent storm events and sea defence construction. An $8 million plan (2020) for rehabilitation and stabilisation ("nourishment"—the British term is re-charging) through the addition of 50–100,000-cubic yards of sand sourced from offshore "borrow" areas will expand the beach by widths between 11 and 46 metres. This will not only buffer it against erosive waves but also increase the amount of "dry" beach at beachgoers' disposal.

L is for Antonie van Leeuwenhoek

A renowned early arenophile—perhaps the original—was self-taught Dutch scientist Antonie van Leeuwenhoek (1632–1723). A successful draper, he developed an interest in glass grinding and learnt how to make his own single-lensed microscopes, with which the pioneering microbiologist discovered bacteria, among other things. Describing foraminifera (single-celled protists) as "little cockles […] no bigger than a coarse sand-grain", van Leeuwenhoek also used his microscopes to enlarge actual grains of sand.[28] In a letter written to the Royal Society (4 December 1703), he included a red chalk drawing of sand grains from the Dutch East Indies as they appeared to him through his microscope. What we really owe to van Leeuwenhoek is an appreciation of the uniqueness of each grain. "I remember I have formerly affirmed of Sand", he wrote in that letter, "that you cannot find in any quantity whatsoever two Particles thereof, that are entirely like each other, and tho perhaps in their first Configuration they might be alike, yet at present they are exceeding different".[29]

M is for Margate

Donkey rides are no more, but north-facing Margate Main Sands in Kent remains perhaps the ultimate Great British "bucket-and-spade" beach (Brighton's is pebbled). The wide sands slope gently into the sea so children can paddle safely; the golden sand is soft and ideal for building castles; deckchairs are for hire; fish and chip shops are just a stone's throw away; ample car parking is available nearby and the railway station (central London is only an hour and a half away) literally opens onto the seafront. Opportunistic seagulls, invigorating breezes and dramatic sunsets that local resident J. M. W. Turner loved to paint complete the scene. Overlooking the beach is the Turner Contemporary gallery, whose first photography exhibition, in the summer of 2019, "Seaside: Photographed", aptly, was all about the beach and beach life.

N is for Nurdle Beach

Widemouth Bay Beach is a broad, mile-and-a-half stretch of proverbial "golden sands" near Bude, north Cornwall. Until a few years ago, Widemouth (pronounced "wid-muth") had a simple reputation: as one of Cornwall's best Atlantic beaches for surfing, bodyboarding and family holidays. Since the weekend of 3–5 February 2017, that reputation has been complicated, and somewhat tarnished.[30] While oil spills have darkened the histories of beaches from Santa Barbara, California (1969) to Barafundle Bay, Pembrokeshire (1996), beaches like Widemouth have been tainted more recently by a tide of tiny plastic objects. "Nurdles" are the pre-production building blocks for nearly every plastic product, from drinking straws to keyboards. A lentil-sized raw plastic pellet, the nurdle is readily transportable, meltable and mouldable. These lightweight, bouncy and floating discs also spill easily during manufacturing and transit, then wash or blow down storm drains designed to intercept bigger and heavier materials.[31] Then they flush out to sea, eventually fetching up on beaches like Widemouth.

According to Fidra, a Scotland-based environmental NGO (named for a small island in the Firth of Forth), an estimated 53 billion nurdles a year "escape" into the UK's marine environment from terrestrial sources (2016).

Many more are spilled directly into the sea during transportation and cargo handling (the UK imports almost as many nurdles as it produces—and each tonne consists of 10 million). Birds and fish ingest these fish-egg resembling "mermaid tears", whose toxicity exacerbates as they absorb persistent organic pollutants such as PCBs from their surroundings. Potentially lethal microbes such as *E. coli* also colonise nurdles, which simply fragment over time under UV light into ever tinier particles.

That weekend in February 2017, in collaboration with Flora & Fauna International, Greenpeace, the Marine Conservation Society and Surfers Against Sewage, Fidra organised the Great Winter Nurdle Hunt Survey (the beach clean-up phenomenon can be traced back to an Ocean Conservancy initiative on the Gulf Coast of Texas in 1986). Six hundred volunteers scoured 279 coastal locations from the Scilly Isles to the Shetlands. They found nurdles at almost three-quarters of the sites sampled. If any of those sites deserves the dubious moniker of Nurdle Beach, it is Widemouth, where 33 volunteers from the Widemouth Taskforce collected the weekend's record number: 127,500 pellets along a 100-metre section of beach. The Great Nurdle Hunt has become an annual event held internationally in February. Nurdle Free Oceans, a Fidra initiative, maintains a team of nurdle hunters that monitors the world's beaches and has organised over 2200 hunts to date, logging the data onto a global map.

Nurdles should not be confused with microplastics. These minute pieces measuring less than 5 mm, often smaller than a grain of sand, represent the mechanical (as distinct from biodegradable) breakdown of larger pieces of plastic into ever more miniscule bits that marine biologist Richard Thompson first named in 2004. This pioneering identification of the scale of the microplastics problem was also based on materials collected from Britain's beaches. Thompson's team compared them to plankton samples collected regularly since the 1960s along sea routes between the north of Scotland and the Shetlands and Iceland respectively.[32] Since then, the amount of often fibrous and brightly coloured plastic "archived among the plankton" in the water column has steadily grown and increasingly matches the microplastic content of the grains gathered from beaches. But there is no equivalent for microplastics to the great nurdle hunts. Citizen science-style beach clean-ups like Widemouth's in 2017 cannot reveal the scale of microplastic pollution as many microparticles are concealed beneath the surface.[33] Scientists are just beginning to understand how the embedding of microplastics within sand is affecting its temperature and how water flows through it. For now, Microplastic Beach is a generic, non-site specific concept.

O is for Omaha Beach

Omaha Beach was the code name for an 8-km stretch of "golden" and firm sanded beach in German-occupied Normandy, northern France, during the Allied Forces' D-Day landings on 6 June 1944 (one of five landing points along an 80-km section of coastline). The US Army's hard-won objective was to secure a beachhead to link up with a British landing to the east and an earlier American landing to the west at Utah Beach. Given that all battle debris was removed long ago, it was thought, until recently, that no physical evidence survived. But when they visited in June 1988, US geologists Earl McBride and Dane Picard (hard-core arenophiles) collected a jar of sand from a point near the high-water mark, close to the War Memorial overlooking the beach. On returning home, though, they effectively put the jar on the shelf. When they finally gave it their full attention in 2011, they were surprised to find significant quantities of angular, rusty metallic "grains"—shrapnel particles—among their sample (4%). This unexpected material record of the landings—how representative the quantity of shrapnel in the sample is of the beach as a whole is unclear—will probably endure the onslaught of chemical corrosion from saltwater and mechanical abrasion from wave action for at least another century.[34] However, these strange, sharp-edged grains will eventually disappear, leaving only regular rather than militarised sand.

P is for Psammophile

Psammophile is not just another term (derived from ancient Greek) for a person who loves sand (see *Arenophile*). It is also a biological term for a plant that thrives in a sandy environment, such as Aleppo pine, sea holly, sea daffodil, Cretan trefoil, marram grass and European searocket.

Q is for Quicksand

The opposite of sand so hard you can ride or drive on it is sand that, despite the appearance of solidity, can quickly liquify and then collapse. Quicksand is fine sand and clay so heavily saturated with water that the usual friction between individual particles is reduced to the point where the substance cannot bear even the slightest weight. At the merest disturbance, water separates from the sand and clay as viscosity (resistance) weakens. Just one per cent of physical stress causes the flow speed of the affected particles to multiply a millionfold. Even a single human foot is hard to pull out, requiring strength equivalent to that needed to lift a medium-sized automobile. That does not mean, though, that complete submergence is just a question of time.[35] Despite what numerous B-movies have suggested over the past 70 years (peaking in the 1960s), sinking in completely is physically impossible.[36]

Research by physicists at the University of Amsterdam (2005) involving laboratory simulations based on a sample from a saltwater lake in northern Iran showed that our bodies—like aluminium beads and plastic toys—are not dense enough.[37] Even those who do die in quicksand get stuck rather than sucked under. In 2015, a man swimming in the San Antonio River, Texas, got caught in quicksand "to the bottom of his buttocks" and, unable to extricate himself, probably died of thirst and exposure, or maybe drowned when the river level temporarily rose after heavy rains upstream (news reports are unclear regarding cause of death). Unless the quicksand you are trapped in is within an intertidal zone subject to big tidal variation, your chances of survival are reasonably good. Resist the urge to fight the quicksand, and you are unlikely to sink above the waist. The particles will eventually re-settle and the buoyancy of the mix of sand, clay and water slowly lift you to the top. But thanks to the enduring pull of films such as *The Mongols* (1961)—in which Anita Ekberg expired in a tank in a studio filled with a metre of sand and water—and *Lawrence of Arabia* (1962), that sobering scientific truth will take a long time to sink in.

R is for the Right Kind of Sand

The only natural resources we use more than sand are water and air. As Beiser observes, sand is the "most important solid substance on earth".[38] Some types of sand are more suitable for human uses than others. Ironically, beach sand is actually the wrong kind for beach volleyball: too hard for injury-free play, it also sticks to sweating skin. Recently, a global specification has been developed that regulates grain size, shape and hardness. The resultant product drains so well that a court is match-ready soon after heavy rains. Sourcing an accessible supply of the right kind of sand for the Olympics and other international tournaments confronts suppliers with a major challenge.[39] As not all sand is commercially usable as "aggregate", whether for beach volleyball courts, coastal fortification or key ingredient for concrete, sand, to quote a United Nations Environment Programme report (2014) is "rarer than one thinks".[40] Most desert sand, for example, has no construction value because the round grains formed by wind erosion bind ineffectively: the sand in the bunkers of Dubai's golf courses is imported from North Carolina, Ontario and Australia.

S is for Sandman

A mythical character in Northern European folklore, the Sandman sprinkles sand (or fine dust) into children's eyes at night to get them to sleep and also to dream, as related in Hans Christian Andersen's story, *Ole Lukøje* (1841), named for the main character, who represents the Sandman. The gritty "sleep" (rheum) we rub from our eyes when we wake up is supposedly the remnant of his nocturnal activity. A more sinister persona features in E. T. A. Hoffmann's story, *Der Sandmann* (1816), in which he tosses sand into the eyes of naughty children who refuse to sleep. Their eyes then fall out and the Sandman takes them to his nest on the Moon, where he feeds them to his own children.

T is for TIGL (Trump International Golf Links)

"Direct encounters with sand", reflects Michael Welland, "are typically at the beach or, more frustratingly, on the golf course".[41] The word links, a synonym for golf course, is derived from linksland, a sandy coastal environment shaped by the wind and characterised by dunes and hollows carpeted with grass kept short by grazing and strong winds (*hlinc* is Old English for "rising ground"). Linkslands line the Atlantic coast of Outer Hebridean islands such as South Uist.[42] This terrain lent itself naturally to golf as well as grazing, and Scotland's renowned links along its North Sea "golf coast" embrace these ingredients. The bunker probably originated as a sandy hollow abraded through skimpy turf by cattle huddled together for shelter. In fact, many of Scotland's early courses required little re-engineering; the design flowed from the lay of the sandy land.

Dune systems are all more or less unstable; parts of the original greens established in the late nineteenth century at Askernish on South Uist had been lost to coastal erosion or buried deep under windblown sand when the course was refurbished (reopening in 2008). One of northwest Europe's finest and largest examples of a dynamic windblown dune system is Foveran Links, Aberdeenshire. Shaped by thousands of years of reconfigurations as the spit and bar complex at the mouth of the Ythan, the sandy beach and the dune belt exchange materials in an ongoing cycle of deposition and removal governed by wind and waves; historically, the dune system moves north at a rate of some ten metres per year. The associated flora of this Site of Special Scientific Interest (SSSI, est. 1984) is rich and complex as highly mobile sand scours down to the water table, uncovering wet sand that plants colonise before themselves becoming covered.[43]

Infamy now complicates Foveran's fame. In the summer of 2012, the Trump International Golf Links controversially opened for business at Menie Links, Balmedie, a site overlapping with a third of the SSSI. Oblivious to what was required to create and maintain the topographical features that attracted him, Trump (2007) hailed proposals to stabilise the dune "dome" by planting marram grass as a tremendous environmental improvement: "dunes can be gone with the wind […] it's a piece of land which is disappearing. It's blowing all over the place".[44] Tarmacadamed paths for golf buggies cross turf laid for greens and fairways over scraped and levelled sands. Stripped of many of its signature features, ecologists and environmentalists believe Foveran Links risks losing its SSSI status.

Whereas other "golf coast" links were laid out among fairly stable sands, Trump's 18-hole course "amidst the Great Dunes of Scotland" (membership brochure) tries to "freeze" sand, to defy a highly restless dune world. Ancient cities have been entombed in sand in the Sahara and Taklamakan deserts, a fate that may also await Trump's project.

U is for Uber Beach Bag

Uber launched its free promotional Beach Bag in selected cities in New Jersey, New York and New England in July 2014. The promotion is now annual, and for the fortunate few, the tote bag (delivered by an Uber driver) comes filled with other branded goodies, such as sunscreen, a skincare gift pack, sunglasses, baseball hat, beach ball and a gift card for a pair of espadrilles.

V is for Volcanic Sand

"Volcanic blend" is not a variety of coffee but a dark sand type that pet owners can buy from the company Komodo to provide a "safe and hygienic terrain" for "desert dwelling reptiles". Black sanded beaches (not the source for "volcanic blend") can be formed by heavy minerals such as magnetite or garnet, but most—and certainly the best known—are volcanic in origin. They are associated with volcanoes' extrusive igneous rocks, comprising massively eroded fragments (mostly) of basaltic lava, which blackens as it cools and firms up. Volcanic sand comes from the erosion of this volcanic terrain's immature rock fragments, which, unlike mature sands formed of weathering resistant quartz, readily decompose. Or black sand can derive from eruptions. Iceland has spectacular black beaches, such as Reynisfjara. The Aleutians, Canaries and Azores also have notable black beaches, but Europe's most famous is probably Ficogrande on Stromboli, a miniature Aeolian island off the northeast coast of Sicily, right under the eponymous volcano that rises 2700 metres out of the Tyrrhenian Sea.

The most renowned beaches of this kind, however, are found on the Hawaiian Islands. Kehana Punalu'u Beach was forged literally overnight when scalding lava met saltwater. Volcanic sand beach is actually a transient feature unless the black material is replenished by further eruptions: in Hawaii, it is illegal to remove black sand. Yet an eruption can jeopardise a volcanic beach in the short term. In 1990, lava flows obliterated Kaimu Beach (aka Black Sand Beach) and also buried the nearby town of Kaimu 15 metres deep; meanwhile, though, a new Kaimu Beach is forming. Hawaii's newest black sand beach was created by an eruption at Pohoiki in the summer of 2018. A lava flow emptied into the sea, forming a mass of black sand that the currents shoved down the coastline to Pohoiki Beach, in Isaac Hale Beach Park, where it currently obstructs the entrance to the boat harbour. Eruptions give black sand and take it away.

Though they reflect the sun more softly, volcanic beaches are also hotter. Black sand absorbs more solar radiation than white or green sanded counterparts. Walking on them barefooted can feel like the proverbial cat on a hot tin roof.

W is for Wet Sand

Though this entry is not about "Wet Sand" (2006), a song by the Red Hot Chili Peppers, like the band, it is rooted in coastal California. Despite state legislation (1976) guaranteeing open entry to the state's 1350-km coastline as a fundamental right, the super-rich and other property owners have tried various means to privatise beaches adjacent to their beachfront properties, such as the deployment of security guards and erection of "no trespassing" signs. But activist groups such as California Coastal Protection Network and the Surfrider Foundation are starting to enjoy success in legal battles that have lasted up to a decade. In 2018, the US Supreme Court turned down a last-ditch appeal by Silicon Valley billionaire Vinod Khosla against an order to unlock an access gate to Martins Beach, a popular surfing and picnicking spot in San Mateo County (the 1976 law requires property owners to permit public access through their land in the absence of alternative forms of access).[45] In 2019, the California Coastal Commission, which administers the 1976 Coastal Act, fined the Ritz-Carlton Hotel in Half Moon Bay (just to the north of Martins Beach) $1.6 million for repeatedly blocking public access.

The legal basis for universal access is the "wet sand" doctrine, which stipulates that the beach is public up to the mean high-water line. Still, drawing a line in the sand between public (wet) and private (dry) will prove increasingly difficult, as rising sea levels move the demarcation line between wet and dry sand—and public and private property–further up the beach.[46]

X is for Xanadu Beach

Xanadu Beach, at Freeport, Grand Bahama, is the white powder sand beach of tourists' Caribbean dreams. Nowadays, though, it is much quieter than it was in the 1980s and 1990s, with an abandoned feel. Over a decade ago, the adjacent Xanadu hotel and resort—a getaway destination for Hollywood stars after US property magnate Howard Hughes bought it in 1972—shut down after sustaining severe hurricane damage in 2004–2005.

Y is for Yemen

War and famine-ravaged Yemen, located on the southern strip of the Arabian Peninsula, is as rich in sand as it is poor in monetary wealth. The country's northeast is dominated by the reddish-orange dunes of the Rub' al Khali desert ("Empty Quarter") that occupies the southern third of the peninsula and represents the world's largest expanse of sand (bigger than mainland France).

Z is for Cra-Z-Sand

Cra-Z-Sand, made by LaRose Industries of Randolph, New Jersey, is a nontoxic and antibacterial substance that never dries out ("amazing creative sand"). It is available in various colours (including sparkling pink and sky blue) for those aged four and above to create "no-mess" sandcastles away from the beach ("Shape It, stack it, slice it!").

Postscript

These alphabet entries are selective as well as eclectic. C could easily be for Concrete (difficult to make without sand) or Coral Springs Beach (Jamaican site of one of the world's most notorious sand heists in 2008). And as I check these entries in late March 2020, as social distancing is enforced on beaches as well as in supermarkets, from Weston super Mare to Sydney, C could also be for Covid-19 beach closure. Likewise, D could stand for Dune; F for Fulgurite[47]; G for Glass, Glass Beach, Golf or Gobi Desert; H for Sand Heist (see Coral Springs beach); M for Marram grass, Sand Mining or Sand Mafia; N for Nourishment (beach sand replenishment); S for Sandstorm, Sahara, Singapore, Silicone or Silicon; W for World islands (Dubai) or White Sands, New Mexico. Throughout, though, I've tried to mix three ingredients: substances, processes/activities and sites. In other words, different kinds of sand (biogenic, volcanic), various things we do to sand (collect, dredge) and a range of specific sandy places. The third element is clearly the most personal. M, for instance, might have been for Marina Beach or P for Poyang, rather than these two sites featuring under Dredging. But I resisted the temptation to choose my boyhood beach at Formby Point, Merseyside—renowned for the prehistoric human and animal footprints preserved in the foreshore mud for millennia but now exposed by sand erosion[48]—as the entry for F.

Despite its eclecticism, selectivity and personalisation, certain themes run through this alphabet from start to finish. Firstly, the sheer variety of sand as a material entity and its diverse locations. Secondly, the range of activities that sand supports and inspires, exploitative, recreational and protective: gathering, removing, collecting, playing, treasuring, replenishing and saving. The final theme is the spectrum of values we attach to sand: functional, commercial, aesthetic, amenity, ludic, scientific and ecological—a spectrum as wide as the beach at Ainsdale, immediately north of Formby, Merseyside, where I learnt to drive.

Notes

1. David Brodsky, *Spanish Vocabulary: An Etymological Approach* (Austin: University of Texas Press, 2008); James Duffy, *Sand of the Arena: A Gladiators of the Empire Novel* (Ithaca: McBooks Press, 2005). Chariot races were also run on

a top surface of sand: 20 centimetres composed the set for the chariot race in the 1959 film, *Ben Hur*, dir. William Wyler (Beverley Hills, CA: MGM). See Eckart Köhne, Cornelia Ewigleben, and Ralph Jackson (eds.), *Gladiators and Caesars: The Power of Spectacle in Ancient Rome* (Berkeley: University of California Press, 2000), 96–98.

2. "Jim's Sand", http://scimuze.com/sand_images/; Mary Hemmerly Hecker Memorial Sand Collection, http://www.maryhemmerlyhecker.org/; Niklas' Sandcollection, http://www.sandcollection.org/; Sandatlas, http://www.san datlas.org/; Sand. World (collection of Daniel Helber), http://www. sand.world/sand-sammlung/index; Something Sand, https://somethingsand.blo gspot.co.uk/; The Virtual Sandbox—Museum of World Sands, http://sand.aja ster.com/; World Atlas of Sands, http://www.sand-atlas.com/en/.

3. Swiss arenophile Marco Bonifazi (who possesses over 8000 samples), co-curated an exhibition, *Le Sable*, at the Natural History Museum in Neuchâtel (2002/2003). An avid current practitioner is Michael Welland, geologist and author of the 2009 *Sand: The Never-Ending Story* (Berkeley: University of California Press), who maintains the lavishly illustrated website, "Through the Sandglass", http://throughthesandglass.typepad.com/.

4. The International Sand Collectors Society, "The hobby", http://sandcollectors. org/the-hobby/; Sand Trading Tips, "Sand Collecting: The Hobby of Sand Collection", http://sandcollecting.net/sand-trading-tips.php.

5. E. Özhan, "The legend of Cleopatra Beach: May It Be True?", *Eurocoast* (1990): 98–103.

6. Amr El-Sammak and Maurice Tucker, "Ooids from Turkey and Egypt in the Eastern Mediterranean and a Love-Story of Antony and Cleopatra", *Facies*, 46 (2002): 217–228.

7. Pascal Peduzzi, *Sand, Rarer Than One thinks* (Sioux Falls, SD: UNEP Global Environmental Alert Service, March 2014), 6; Vince Beiser, *The World in a Grain: The Story of Sand and How It Transformed Civilization* (New York: Riverhead, 2018), 181–197.

8. David Owen, "The end of sand", *New Yorker* (29 May 2017), 32.

9. Beiser, *World in a Grain*, 2.

10. Peduzzi, *Sand, Rarer Than One Thinks*, 5; Vince Beiser, "Sand Mining: The Global Environmental Crisis You've Probably Never Heard Of", *Guardian* (27 February 2017), https://www.theguardian.com/cities/2017/feb/27/sand-mining-global-environmental-crisis-never-heard.

11. Beiser, *World in a Grain*, 11–12. CEMEX's operations are being phased out, with total shutdown by the end of 2020: Paul Rogers, "Controversial Beach-front Sand Mining Operation Along Monterey Bay to Close", *The Mercury News* (27 June 2017), https://www.mercurynews.com/2017/06/27/controver sial-beachfront-sand-mining-operation-along-monterey-bay-to-close/.

12. Jesus Martinez-Frias, "Lanzarote: Mars on Earth", in *Lanzarote and Chinijo Islands Geopark: From Earth to Space*, eds. Elena Mateo, Jesus Martinez-Frias, and Juana Vegas (New York: Springer International, 2019), 143–148.

13. Matthew Chojnacki et al., "Boundary Condition Controls on the High-SAND-Flux Regions of Mars", *Geology*, 27, no. 5 (2019): 427–430.
14. A. A. Mills, S. Day, and S. Parkes, "Mechanics of the Sandglass", *European Journal of Physics*, 17 (1996): 98–99.
15. C. B. Drover, P. A. Sabine, C. Tyler, and P. G. Coole, "Sand-Glass 'Sand': Historical, Analytical and Practical", *Antiquarian Horology*, 3 (1960): 62–72.
16. R. T. Balmer, "The Operation of Sand Clocks and Their Medieval Development", *Technology and Culture*, 19 (1978): 615–632.
17. Allan Dwan (dir.), *Sands of Iwo Jima* (California: Universal Studios, 1949); Clint Eastwood (dir.), *Letters from Iwo Jima* (Burbank, CA: Warner Bros, 2006).
18. Larry Smith, *Iwo Jima: World War II Veterans Remember the Greatest Battle of the Pacific* (New York: W. W. Norton, 2008), 326.
19. Kyle Munson, "70 Years Later, 20 Ounces of Iwo Jima Sand" (25 March 2015), https://eu.desmoinesregister.com/story/news/local/kyle-munson/2015/03/24/frank-pontisso-world-war-ii-veteran-iwo-jima-trip-return/70405198/.
20. *Daily Mail* (24 July 1933), 18.
21. Mary Lyons, *The Story of Jaywick Sands Estate* (Cheltenham: History Press, 2005).
22. *Daily Mail* (15 May 1934), 20.
23. Documentary films include *Jaywick: A Diamond in the Rough*, dir. Penelope Read and Heather Thompson (2016), and *Benefits by the Sea: Jaywick* (Spun Gold for Channel 5, ongoing since 2016).
24. Negative press coverage of Jaywick Sands peaked in the autumn of 2018 when a photo of a dilapidated street (Essex Street) featured anonymously—splashed with the slogan "Only YOU can stop this from becoming REALITY!"—in a bizarre "attack" commercial made by a Republican candidate for the US mid-term congressional elections. The intended message was that the way to prevent American communities from deteriorating into somewhere like Jaywick was to vote against the Democratic party candidate: Helena Horton, "Essex Seaside Village Outraged After Being Used in Pro-Trump Campaign Advert", *The Telegraph* (31 October 2018).
25. Michael Rouse, *Clacton-on-Sea Through Time* (Stroud: Amberley Publishing, 2009); Fred Gray, *Designing the Seaside: Architecture, Society and Nature* (London: Reaktion, 2006), 288.
26. Roger Deakin, *Waterlog: A Swimmer's Journey through Britain* (London: Chatto & Windus, 1999), 263.
27. Christine Lee, *The Midwife's Sister: The Story of Call the Midwife's Jennifer Worth by Her sister Christine* (London, Verso, 2015), 8. In 1952, a boy had to be dug out when the sides of a hole he was digging collapsed: "Boy, 9, buried on beach: 4ft hole caves in", *Daily Mail* (1 September 1952), 5.
28. Michael Welland, *Sand: The Never-Ending Story* (Berkeley: University of California Press, 2009), 29.

29. Antonie van Leeuwenhoek, "Part of a Letter from Mr Anthony van Leuwenhoek, F. R. S., Concerning the Figures of Sand", *Philosophical Transactions of the Royal Society of London*, 24, no. 289 (1 January 1705), 1537.
30. BBC, "Plastic 'Nurdles' Found Littering UK Beaches" (17 February 2017), https://www.bbc.co.uk/news/uk-39001011.
31. Claire Gwinnett, "The Major Source of Plastic Pollution You've Probably Never Heard Of", *The Conversation* (14 February 2019), https://theconversation.com/the-major-source-of-ocean-plastic-pollution-youve-probably-never-heard-of-111687.
32. Richard C. Thompson et al., "Lost at Sea: Where Is All the Plastic?", *Science*, 304 (7 May 2004): 838.
33. Laura Parker, "Beach Cleanups Are Missing Millions of Pieces of Plastic" (16 May 2019), https://www.nationalgeographic.co.uk/environment/2019/05/beach-cleanups-are-missing-millions-pieces-plastic [based on Jennifer Lavers, *Scientific Reports*, 16 May 2019].
34. Earl F. McBride and M. Dane Picard, "Shrapnel in Omaha Beach Sand", *Sedimentary Record*, 9, no. 3 (September 2011): 4–7.
35. Claudia Hammond, "Can Quicksand Really Suck You to Your Death?", *BBC Future* (29 March 2016), http://www.bbc.com/future/story/20160323-can-quicksand-really-suck-you-to-your-death.
36. Daniel Engber, "Terra Infirma: The Rise and Fall of Quicksand", *Slate* (23 August 2010), http://www.slate.com/articles/health_and_science/science/2010/08/terra_infirma.single.htm.
37. A. Khaldoun, E. Eiser, G. H. Wegdam, and Daniel Bonn, "Liquefaction of Quicksand Under Stress", *Nature*, 437 (29 September 2005): 635.
38. Beiser, *World in a Grain*, 5, 2.
39. David Owen, "The End of Sand", *New Yorker* (29 May 2017), 28–33.
40. Peduzzi, *Sand, Rarer Than One Thinks*.
41. Michael Welland, *Sand: A Journey Through Science and the Imagination* (Oxford: Oxford University Press, 2009), xiii.
42. David Owen, "The Ghost Course: Links to the Past on a Scottish Island", *New Yorker* (20 April 2009), 36.
43. Scottish Natural Heritage (Aberdeen), "Foveran Links Site of Special Scientific Interest: Management Statement" (1 February 2002).
44. Peter Popham, "Tee'd Off: The Residents of Foveran Links Speak Out About Donald Trump's Golf Project", *Independent* (14 July 2012); Scottish Wildlife Trust, "Our objection to the Trump Development" (Edinburgh, 2009); Anthony Baxter (dir.), *You've Been Trumped* (Montrose Pictures, 2011).
45. Rosanna Xia, "U.S. Supreme Court Declines to Take Martins Beach Case—A Win for California's Landmark Coastal Access Law", *Los Angeles Times* (1 October 2018); Gregory S. Alexander, "The Choppy Waters of Beach Ownership: A Case Study", *OUPblog* (23 April 2018), https://blog.oup.com/2018/04/choppy-waters-beach-ownership-case-study/.

46. Dan Moffett, "Along the Coast: Public or Private? Line in the Sand Unclear", *The Coastal Star* (2 May 2018), https://thecoastalstar.com/profiles/blogs/along-the-coast-public-or-private-line-in-sand-unclear.

47. A fulgurite is a fragile, hollow shaft of glass up to 15 metres long that is created when a lightning bolt strikes the desert, heating the sand to temperatures higher than 10,000 degrees C.

48. Gordon Roberts, *The Lost World of Formby Point: Footprints on the Prehistoric Landscape, 5000 BC to 100 BC* (Formby: Alt Press for Formby Civic Society, 2014).

4

An Englishwoman's Home Is Her Castle: Social Morphologies and Coastal Formations

Sefryn Penrose

I bear with me an ocean shell,

Its sounds to me are dear, —
Oh! like an old familiar spell,

It murmurs in my ear;
Even in exile, I rejoice

In some still spot to be,
And greet its low mysterious voice,

The language of the sea.
Maria Abdy, "My Home is By the Sea".[1]

* * *

I'm sitting at the kitchen table with my mum, Angela, old photo albums, the AA Guide to Britain's Coast, and a bottle of wine. I've asked her to tell me about her childhood seaside holidays. We're going to start at – where are we going to start? Margate?
We have to start near South London where my parents lived. So just after the war they didn't think much beyond the Thames Estuary and Kent, so

S. Penrose (✉)
Plymouth, UK

© The Author(s) 2020
J. Carruthers and N. Dakkak (eds.), *Sandscapes*,
https://doi.org/10.1007/978-3-030-44780-9_4

Fig. 4.1 Paddling with Eric at Ryde, 11 months (left, July 1948); and Joan, at Margate (July 1949) (*Source* © Sefryn Penrose)

Margate, Westgate Bay. I think that's where we went when I was a toddler and it was just sea and sand. I'm assuming we stayed in a boarding house. I think it's what they needed. You just got out of town and you had perfect freedom and not a lot of pressure on what you looked like or what you did.
In a boarding house, you had to be out during the day?
Yes and there are photographs of my father—and my mother—and they're both wearing Gaberdines. You're just out there with a headscarf and a mac on and a couple of kids in pushchairs and lots and lots of sand—you just went on the sand. There are endless pictures of sand: sandcastles, sitting on the sand, or I guess the prom.
Did you go in the water? Did they go in the water?
My father certainly went in the water and he must have introduced me to the water very early and we would have paddled—I mean paddled—you paddled. It was a joy. It was absolutely wonderful. It's what they looked forward to for months and months and months and presumably saved up for (Fig. 4.1).

* * *

Maria Abdy (1797–1867) picked up the shell again. In "The Song of the Sea-Shell", she has it speak,

I am bound by mysterious links to the ocean,

And no language is mine but the sounds of the sea.[2]

One of my earliest memories is sitting on the bottom of a swimming pool and looking up, feeling very calm in an early acknowledgement of a diluvian truth. And while I remember my brother rescuing me—raked-moon-face—I'm told it was my mother who actually hauled me out. An inauspicious start for a sea-creature, but never-the-less… I know there are some people who don't like the sea. People who prefer mountains to beaches, but we are not they.

Maria Abdy died in Margate. The coroner poetically inscribed her death as due to "the decay of nature". Her neighbour was a surgeon at the Royal Sea Bathing Hospital overlooking the next bay along—set up for the "scrofulous poor" of the East End. A different East End demographic to the exiles that have gifted it its new nickname of Shoreditch-on-Sea.

* * *

And if you were in a boarding house, Mrs Brown would make your breakfast and your dinner.
Certainly, you had breakfast and you went back for tea or supper and your evening meal—it wasn't called dinner. And I don't know what you did at lunchtime. We would have moved in my first few years from a boarding house up to a hotel. A family hotel.
What were they like? You would have all been in one room.
My parents would have presumably booked a family room for two weeks—the best two weeks of their year.
That must have been really expensive.
I think for them it was huge. They didn't go away at any other time in the year. My mother certainly never would have spent any time in a hotel.

* * *

But the large seaside resort was already on the downward slide in the immediate post-war period. Keith Parry, historian of the Lancashire coast, suggests landlady stagnation in Blackpool as early as the 1930s with no new boarding houses opening up.[3] The Paid Holidays Act had only come into force in 1938, though, and while white-collar workers more or less had their holiday habits in place before then, industrial workers did not. In those early post-war years, my grandparents moved to the North West—Eric following a job from apprenticeship to partner as a quantity surveyor—moving into the middle classes and accelerating their mobility through motoring into Manchester's growth years.

* * *

Right. Margate, Ramsgate, Folkestone, Hastings. Eastbourne, Brighton, Bognor Regis.
Well I didn't realise that I'd more or less had a holiday in Bognor Regis but in fact it was Felpham which is to the east of Bognor. It has old, cobbled streets, ancient houses. There are pictures of me on the beach at Felpham.
August 1952. We're looking at a photograph of David and Angela in 1952, Felpham…. David looks delighted, atop a mound of sand…
I'm looking absolutely fabulous in my jacket.
Both of you are looking particularly well dressed. For an August.
Yes, my brother's got three layers on so it must have been a bit chilly.

* * *

Catherine Blake, cognisant of the fashion for bathing at Bognor took to the water at Felpham: "My Wife & Sister are both well, courting Neptune for an embrace…" wrote William in between his lines on Milton.[4] But after three years, the miasma of London seemed preferable to the "unhealthiness of the place", its agues, "Neptune's terrors".[5] Perhaps she was a seasonal bather, unused to the storm surges and whipping, wet, onshore winds. As her husband put it, Satan "howld round upon the sea".[6] William Hayley (for whom, while in Felpham, Blake was engraving illustrations) courted it: "he would thus plunge at periods which, to our softer ideas, seem positively inhumane. October bathing, for instance, was nothing to him".[7]

Another photograph, an albumen carte-de-visite of 1901, in the National Portrait Gallery's photographic collection. Mary Wheatland wears a straw hat with her name printed on a ribbon around it, over "Teacher of Swimming". She has a bathing dress draped over her arm. In others, she wears medals— received for bravery, life-saving. Her sea-smooth face is beach-tanned. In another she stands in front of the giant cartwheels of the bathing machines that she ran, the muse of the local photographer.

The gap between Felpham and Bognor was bridged in 1960 by Butlin's Holiday Camp.

* * *

Fig. 4.2 The Mickey Mouse tin bucket, in Exmouth, 1949; Felpham, 1952 (*Source* © Sefryn Penrose)

And what's on that bucket?
IT'S A MICKY MOUSE TIN BUCKET! When did tin buckets go out of fashion?
You are against a wooden fence.
It's a breakwater!
Oh, you're against the groynes, because it's so windy and cold in Felpham in 1952. You're almost five and David is two.

* * *

Bognor takes its name from a woman—Bucge—with the suffix *ōra*, a flat-topped hill. Ann Cole, writing for the *Journal of the English Place Name Society* suggests that the *ōra* must lie between the observer and a hill—in this case the Sussex Downs. The sailor approaches Bucge's settlement from the sea (Fig. 4.2).[8]

* * *

So we're going around the coast. Selsey. So later your aunt Phyl moved to Selsey. There's a picture of Sidlesham –

The Crab & Lobster…
…And a penny farthing. Joan and Eric visited Phyl and Dick every year. That would have been one of my last family holidays with the family. I was fourteen or fifteen.

* * *

Talking of groynes, the Manhood Peninsular (you don't need a map or an Anglo-Saxon specialist to understand the nomenclature here) and its eroding shingle defences is one of those south-coast sites that I've spent grey days at, identifying archaeological potential while the tide tests my safety boots and licks the lips of old sluices, shingle banks, timber groynes, the hundreds of things put in place to try and control its ingress. Violet Hunt, feminist author, outpacing her uneasy relationship with Ford Maddox Ford, sees them too:

> I wandered round one or other of the three beaches of Selsey, conscious of a drone of the sad sea line on the two other sides, pacing on the edge, without looking where I was going, of the octagonal tanks that represented the efforts of the inhabitants of the peninsula to keep the sea in here, let it out there…[9]

The old maps show lost farms, lost plotlands, lost lighthouses and lifeboat stations. You can't stop the sea. Perhaps that's what drove the women of Sidlesham to unrepentant recusancy. The Churchwardens' Presentments of Chichester report a married woman "who scorns our church" standing in the churchyard during services urging the congregation to find other things to do.[10] At least she got that far. A little earlier, in 1608, John Smith was brought before the church to explain his absence on mid-Lent Sunday (he'd been in the pub): "his wife being great with child desired to eat oysters".[11] Selsey, abundant with molluscs at the time I suppose. I noted the remains of the old oyster beds at Medmerry, gone to the tide in the early twentieth century, then went for crab sandwiches at the Crab & Lobster, where my great aunt ate hers. And perhaps Mr and Mrs Smith, and Violet and Ford too—their "little contraband crabs".[12]

* * *

Ok. Lymington? Isle of Wight. Cowes?
No, I was never in the Cowes set. But I'm a bit baffled because I have stayed on the Isle of Wight. I have. I've *been* to Ventnor. I don't know when. I just remember the sand. Do you remember the sand?
I've never been to the Isle of Wight.

No but they sell all these things. All these lighthouses with the different coloured sands from around there. And chines! They're deep river valleys into the… But you only get them on the Isle of Wight. I went to the Isle of Wight. Shanklin Chine! We stayed in Shanklin. We had a holiday in Shanklin.
How old were you when you had a holiday in Shanklin?
I must have been really little. It's absolutely beautiful isn't it. Possibly I do remember a tiny bit of that but it's only just come back to me.
"Shady dell. Ferns and mosses cling to the rock walls and tall trees from leafy sunshade above the deep cleft of Shanklin Chine".[13]
That word just came to me…
…Didn't it?
…it is a deep cleft you see and it's a word I don't think that you find anywhere else.

* * *

Packing up our house. My mother, widowed five years, decided to move from Cornwall to Plymouth, to be in a town, overlooking the sea. I fill a box with the bits and pieces of childhood. Beswick animals, keyrings and bookmarks. A crab-shaped badge from Cromer. A small jar of sand with a gnome's hat that I bought from a kiosk on my first school trip as a Norfolk schoolchild to Overstrand. I can remember the woman in the kiosk. A gatekeeper of seaside treasure. Directing me to the right goods for the pennies in my hand.

"A woman came forth from a cottage to unlock a gate through which we must pass to go up the Chine", Catharine Sedgewick, a New England novelist, an independent woman, wrote home from her Isle of Wight trip in *Letters from Abroad to Kindred at Home*. "K. says the beauties of Nature are as jealously locked up here as the beauties of a harem. It is the old truth, necessity teaches economy; whatever can be made a source of revenue is so made, and the old women and children are tax-gatherers. At every step some new object or usage starts up before us; and it strikes us the more because the people are speaking our own language, and are essentially like our own".[14]

In her bitchily gleeful *My Recollections*, Adeline, Countess of Cardigan (1824-1915), recalls the good-natured adulterous yacht-hopping that occurred in Cowes among the yacht-owning classes, before "the shadow of the American billionaire and the knighted plutocrat" fell, before both aristocrat and keeper of the chine required the stranger to pay up.[15]

* * *

Poole?

No. Except you know about Brownsea Island and the foundation of the Boy Scouts don't you?

No. Did you go to Brownsea Island? I don't know, no.

Sorry, it's just part of my… being brought up as a Girl Guide, you know about Brownsea Island.

* * *

Agnes Baden-Powell was asked by her brother, Robert, to help out giving the renegade girl scouts their own movement after they crashed the 1909 Crystal Palace rally in south London. They'd been signing up without giving their first names. I don't know where the first UK guide camp was held. "Early camping in Britain varied considerably by location and parental concern – some camps were merely overnights in donated houses or barns – others were tent camps. The first documented guide camps appeared to have occurred in 1910".[16] By 1911, the Canadian Girl Guides had held theirs on the banks of the Missinnihe.

* * *

Weymouth? Lyme Regis?

No, but we know all the literature.

Oh yes, that's where they go isn't it, to season.

Yes, they fall off.

Do they fall off?

Yes, they collapse and fall off regularly in Lyme Regis.

* * *

My mother finds the relevant selection from the copy of Jane Austen's *Persuasion* that was my other grandmother's—a great Californian swimmer:

> There was too much wind to make the high part of the new Cobb pleasant for the ladies, and they agreed to get down the steps to the lower, and all were contented to pass quietly and carefully down the steep flight, excepting Louisa; she must be jumped down them by Captain Wentworth. In all their walks, he had had to jump her from the stiles; the sensation was delightful to her. The hardness of the pavement for her feet, made him less willing upon the present occasion; he did it, however. She was safely down, and instantly, to show her enjoyment, ran up the steps to be jumped down again. He advised her against it, thought the jar too great; but no, he reasoned and talked in vain, she smiled and said, "I am determined I will:" he put out his hands; she was

too precipitate by half a second, she fell on the pavement on the Lower Cobb, and was taken up lifeless![17]

* * *

Exmouth. Torbay. Torquay?
No. But here we come to a completely different set of ideas in that Sidmouth is where Sheila loved to go, and Eric and Joan went with Sheila—Eric and Joan being my father and mother, and Sheila being my mother's sister, but being widowed. They went there three years as a threesome in their mature years. It's the sort of place retired people go, but it's not part of my history.

* * *

On a stormy October night in the mid-2000s, we sought refuge in off-season Sidmouth. Easier said than done. The hotels were closed to the unexpected visitor. Dimly lit hotel signs in windscreen wiper intervals promised Agatha Christie Art Deco but pvc doors refused to open. Nevertheless, Sidmouth's wealth of stucco ensures that it remains the backdrop to the high-camp comic drama of Christie TV adaptations. We retreated to the first unpromising 1920s offering we could find that was open, at the top of the town, into its damp chintz. In the morning, our dreams of someone climbing in and out the windows tallied. The uninterested hotelier said that we weren't the first to feel the spectral in the room. My grandmother, a mystery lover, and a Christie connoisseur—she took us to see *The Mousetrap*, and passed on her copies of well-thumbed Miss Marples—and an amateur dramatist would have enjoyed embroidering the tale.

Agatha Christie was born a little to the west, in Torquay. She spent her childhood summers at the women's—ladies—bathing beach at Beacon Cove. In her autobiography, she recalls the segregated bathing of her youth, the ancient, irascible man whose job it was to let the bathing machines up and down into the water. The male bathers, she writes, were further around the bay, "in their scanty triangles", while the women wore the "good old short-skirted British bathing dress of frilled alpaca".[18] But a couple of miles along the bay, at Meadfoot beach, she would *swim*.[19] It's an unlaboured semantic difference that Brits don't seem to use anymore. My Swedish friend Anna says, on the verge of stepping into Plymouth's energetic ocean on the second day of the year, that Swedes still differentiate between *swimming* and a gentler form of submersion, and it's the latter she has in mind. *Bada* rather than *simma*. So we bathe—or more correctly, I suppose, are bathed—for the sea does most of the work and there's no point fighting it. It will bring us back in eventually.

bathe, v.
To take a bath, to plunge or immerse oneself in water or other liquid, so as to enjoy its influence; in earlier usage also, to lie or remain so immersed, to bask.

1667 J. Milton *Paradise Lost* ii. 660 Vex'd Scylla bathing in the Sea.[20]

When she wrote her autobiography, Christie was living at Greenway, a beautiful Queen Anne house overlooking the Dart estuary. The grounds slope steeply down to a slipway and stony beach where a ferry runs the short distance to Dittisham. She gave up swimming due to her rheumatism and difficulty getting in—and out—of the water, but by then she had swum and surfed in many seas from the Levant to Hawaii.

In 1966, Joan and Eric took a holiday cottage in Dartmouth, just down the river. Andrew sailed his model yacht. They spent the hot days on Blackpool Sands. In 1404, the local inhabitants had fought off a Breton invasion. "According to an English chronicler, the Devon women joined in the slaughter with their menfolk" writes historian, Norman Longmate.[21] They used shingle projectiles in place of arrows. The violence of the English seaside then, a site of entrenchment as well as freedom, of attack as well as acceptance. Among other signs, it was the continued discovery of bones and shrapnel on Blackpool Sands, and its neighbour, Slapton Sands, that eventually led a local man to pull a World War II tank from the sea onto the beach in commemoration of the 946 American troops who were killed by German E-boats and friendly fire in 1942 during a training exercise for the Normandy landings.

* * *

Dartmouth. Salcombe?
The only time I went to Salcombe was that holiday with Judy, my cousin. Another night we could spend an hour talking about how I drove the combi down and put diesel in it.
I have a strong memory of waking up… you and dad had gone out for dinner and in the morning I woke up to find crab claws on my bedside table and I thought it was magical. Because they still had the little thing that you can work the claws. You'd put them there… I mean you must have been shitfaced but it was hilarious…
You didn't have a fright?
No, I thought it was magical, a magical thing.

* * *

Among the ghosts are those of the Devon seafarers—those boys that both my mother and I grew up believing to be heroes—Drake, Raleigh, Hawkins. Those boys that sought gold and spices, but enslaved and slaughtered on the way. Devon took to the seas for the fish and when the State began to borrow the men for the Navy, the women went out closer to home for the fruits of the sea. Who came up with Raleigh's cordial recipe that, according to Charles Kingsley, cured Anne of Denmark, wife of James I, and might have won him back his freedom?

B. Zedoary and Saffron, each ½ lb.
Distilled water 3 pints.
Macerate, etc., and reduce to 1½ pint.
Compound powder of crabs' claws 16 oz.
Cinnamon and Nutmegs 2 oz.
Cloves 1 oz.
Cardamom seeds ½ oz.
Double refined sugar 2 lb.
Make a confection.[22]

"Tis Prester John's herb", John Buchan has him say, "but a distillation of sugar and saffron, half made from an old wife's receipt in Devon, and half made from a device of the Indians of Hispaniola. Your quack will mumble charms over it and add noisome things to give it mystery but 'tis only an old wives' posset".[23]

* * *

You never went to those southern Cornish places. Looe, or Polperro, or Fowey.
No. The only time I spent any time on this southern coast was bizarrely in my final year at school when five of us rented a flat above a shop in Carlyon Bay and my mother had to stand guarantor. So, Dallas and Viv, and Felicity, and Frankie and I went down and Jen joined us and we stayed—it was the most bizarre place—and though it was beautiful scenery we were about half a mile inland in a row of shops, above a Post Office, and the flat was, y'know, nothing. But we did see a lot of Cornwall and that was the first time I saw Fowey and I loved Fowey. Jen came down and we actually took a boat across the river.
You were 16. 17.
We were young people and we didn't know what we were supposed to do. But it was quite enjoyable and we met some lads and they all came back one night. And I dunno, a few months later, I was talking to my mother and I

must have been recounting this and she got very upset because apparently she'd signed something that said we would never take any other people into the flat but we'd all had a bit of a riotous evening...

A riotous evening – did you go to the pub?

We did. And I remember this guy saying to me in this pub—a very old person saying—I spent fifty years learning how to kill a pheasant in fifty different ways and now there ain't no pheasants around.

How did you get to Cornwall? Train? You must have got the train from Manchester to... um... St Austell.

But we all got there. How did we all get there?

You all went on the train?

We met Frankie at the station. She was wearing unbelievable sort of white leather, white plastic... things—there's some photos of that somewhere.

And you met a young man? Mevagissey!

This is never ever going to go...

Cross my 'eart.

Right. So we went to Mevagissey. We were overcome by Mevagissey. If you've never been and you've only lived in suburbs, it's really lovely. And I met this young chap and we did get on awfully well. And all these people came to our flat. Anyway, the point is I remember saying to my mother—it must have been first year university as I must have had a boyfriend at the time—"Is it possible to love two people?" And she was very, very critical and said "absolutely not, you couldn't possibly..." Cos I fancied this... I can't remember his name. I just remember he came from Chipping...

...Chipping Norton, Chipping Camden...

...Chipping somewhere...

It must happen all the time...

...holiday romance... So, you said to Grandma is it possible to love more than one person and she said no.

Well, you know, I just didn't get much direction.

What did she say?

No. You can't love two people at once.

So, what she should have said is, yes, there's a thing called polyamory....

Yes! Looking back on it though I think she's right, because then I met Perran and then I couldn't love anyone else...

Anyway Carlyon Bay. Was it your first autonomous holiday?

No! No! Because we'd gone to France!

The same gang?

No, this is the bizarre thing because for two years, Jen and I had gone to France, and compared to that, we were so constrained in Carlyon Bay.

There's a photograph of a young man, who's very attractive. And what was his name?

Dean!

And he was Algerian and you met in Cassis and you communicated for some months via *the written word.*

I thought he was really lovely.

And did you ask your mother... No because you only loved Dean... at that time.

At that time there was nobody else in my life but Dean. It was Sixth Form.

And is this by the seaside?

Yes – Cassis is the most beautiful seaside in the world. It's now totally overrun but then it was lovely. My friend Jennifer and I convinced our parents that we would learn French better if we went to France, so the first year we went, we just bought a rail ticket and we just stuck with the northern bits of France and youth hostels. All on a train. The second year we did it, we went down to the south of France because ... for the simple reason that we were going to do... Tartarin de Tarascon.

What?

Tartarin de Tarascon. Come on Sefryn, you've heard of...

Tartarin to Tarascon. Ok. No.

...in his moulin, which is in the...

Whose moulin? Dean's moulin?

Tartarin's moulin. Tartarin is a person. He comes from Tarascon.

Qui est Tartarin?

Il est un homme qui habite Tarascon.

Il habite dans un moulin à Tarascon? When did he habite à Tarascon dans un moulin?

Well, in the century before the last century but we really did – genuinely – we thought we'd improve our A-levels if we went down to Tarascon to see where Tartarin lived.

Ok. I'm sure that it was true. That you did improve...

Well it didn't really do much for ... anyway... We went all the way down to the south of France. We went to Marseilles. And we went to Tarascon. And then we went along the coast to Cassis – which is where I had this adventure with Dean – and actually what I haven't mentioned here is Jen and a guy in....

Which guy?

Well, we never saw him again.

Hahaha

It was really sad.

Did they ever...

…No they didn't – it was awful – we were youth hostelling – you had to book youth hostels by writing to them. Anyway, I've never seen anything like it. We came off the beach, and this guy said to Jen, "I think I'm burnt, could you rub this stuff into me?"

Aha! Was he French?

No, he was English. And she rubbed this stuff into him. I suppose what I'm saying is I think it was the first time either of us ever thought….

…chemistry….

Thought… Wow! But I don't think anything really happened. We were all so fucking innocent, it's not true.

Where did you meet Dean?

This wonderful place. When I think about it, it's the most wonderful place I've ever been on earth, the youth hostel in Cassis when I was 16. The Mediterranean. This is the south of France before – I don't know – before anybody comes along and – before any regulation, mass tourism. We were the youngest, the most innocent, but it didn't matter.

Where is the youth hostel in relation to the beach?

There is no beach in Cassis. It's called the calanque – rocky inlets and each of the rocky inlets has a teeny weeny beach in the bottom so can you imagine anything better? It's not a great long beach with burnt bodies, it's teeny weeny beaches at the end of little creeks and you have to climb down. The youth hostel was right above and the roof was falling down and there were beautiful beams and it was really lovely. And actually I do wish that my life had gone straight on from there – it was my total perfect life – and this was 1966.

So you persuaded your mother and your father to let you go to Cassis, to learn French.

In retrospect I'm amazed at what they allowed. There was this moment… I bought myself a pair of jeans and for some bizarre reason I cut out the zip and put gold rings instead and a leather thong to join the rings up and as my father put me on the coach he noticed.

You must have seen it in a magazine.

I'd taken the zip out and put gold rings in and wound them up with leather thongs and he just thought – "she's on her way" – he didn't like it but he thought – "she's on her way".

You got the coach from Manchester –

That's right, down to Victoria.

Where you got the boat train.

* * *

That trip to Dartmouth was 1966, before Angela set off for France. She bought her mother (Joan records in her diary) a sea urchin shell, and on the way home, she hung out the window of the car (Ford Zephyr, with newly fixed clutch), "struggling to get reception on the radio". They watched extra time at a hotel in Hungerford. "England beat W. Germany 4/2. Very full of suspense" records Joan.

* * *

St Austell ... Falmouth ...
...Perranporth! You went to Perranporth with your family. 1964.
My father was collecting a new Ford Zephyr car in order to undertake this holiday and the tension was unbelievable because the car hadn't been finished, hadn't been delivered, so if you're going down from Cheshire to Cornwall and I stand to be corrected, but I don't think the motorway existed...

* * *

Joan recorded in her diary the times of their journeys, stops, and mileage. On the way back from Perranporth (leaving 11.35 am), they broke their journey in Minehead (arriving 7.00 pm), setting off the next day at 9.50, to arrive back home at 4.15. The following year, 1965, Joan pronounced, "motorways – boring but useful in shortening journeys" (and a few days later, "car broke down on the way to Woolacombe"). In 1967: "Left 10.30am. Sunny now. Brief coffee stop. Hills outside Ashbourne, 11.50/12. 12.50 got onto M1 (beyond Derby). Picnic lunch at Newport Pagnell, 1.45 til about 2.30. No doubt - you do get through the miles on the motorway. 3.15 – end of M1. Marvellous journey in spite of distance! Got to Selsey at 6.30. Drink in Albion!"

* * *

We got this car and it was absolutely jinxed—it was a green Ford Zephyr and my father wanted to be thrilled with it—it was a beautiful car—but he forever afterwards said it was the Friday afternoon conveyor belt—the Ford guys going off shift... even the door handles didn't work. This poor man at the peak of his career wanted this great big beautiful shiny green Ford Zephyr. We went down to Cornwall and we stayed in a hotel overlooking the beach in Perranporth. And on the beach—as there still is—are Christian groups— these lovely very attractive young people jumping up and down trying to convert the youngsters. I was with my brother Andrew on the beach when a very distinct lady came up to me and said YOU ARE JOAN PEARSON'S DAUGHTER and this was a friend of Joan's from thirty or forty years' earlier and her youth club in Malden.

She was one of the recruiters on Perranporth beach?
Yes. She was with them. She was of the Christian ethic.
Although I've heard this story many times before I don't think I've ever heard that.
Yes. I think it's rubbish. I think I just made that up.
Evangelism on the beach. She recognised your mother through you. Were your parents pleased to see her?
Oh yes. They were thrilled. And they kept in touch. And Andrew kept in touch.

* * *

August 23, 1964, Joan writes: My birthday (43). The colour of the sea is wonderful. Got settled on the beach. Very strange. Met Yvonne Craig (Conway!). Haven't seen her for <u>years</u> & she saw a resemblance to me in Angela & we were all very surprised. Got rather windy and cooler in the afternoon but E. & Angela had a very good bathe.

Yvonne wasn't one of the beach evangelists, but Angela was. Joan records that she went out with CSSM (Children's Special Service Mission) on their beach missions. She says she must have met them on the beach and been interested in young people her own age. I'd never heard of the beach mission before we talked about it, but the day following our seaside chat, we went down to Looe, and there they were, merry pretty people with a little stall, handing out leaflets. CSSM started in 1868 in Llandudno when the founder, Joseph Spiers, wrote "God is Love" in the sand and realised how many children were captive audiences (Fig. 4.3).

* * *

From Perranporth we're going up the coast. Round the headland into Devon. Bideford, no, Barnstaple no. But Woolacombe.
Woolacombe is the most beautiful beach. At the top are these extraordinary rock formations. They zoom out into the sand and I love them. But. I never really felt at home in Woolacombe. I think I'd got too old by then for family holidays.
I'm just consulting the photograph album to see you in Woolacombe.
I think there's just too much put upon people who are trying to understand sand and sea to be at Woolacombe. It's just beyond belief.
Can you expand?
I can't. I'm very happy with Hunstanton or Perranporth.
There's sand at Woolacombe.
Yeah, but it's really fine.

Fig. 4.3 On the sands at Margate, 1948; donkey ride at Exmouth, 1949; sand-car, Skegness, 1955 (*Source* © Sefryn Penrose)

We have a picture of you at Woolacombe. You say you're too old but you're not that old. You're 11!
I think Woolacombe demands that you have an internal...
Monologue...
Woolacombe just ruins the idea of the seaside. I mean the bucket and spade is just lost in Woolacombe.
Because of the rocks?
Because it's the way the rocks zoom out.... it's just really worrying. I think the problem was my father had given me the idea that you can have this easy time with sand and sea.
This is what I'm trying to get at.
What are you trying to get at? You're just trying to get your mother totally pissed.
No – I'm trying to get at the relationship between us and sand and sea. David and Angela, Woolacombe, 1958. It's true to say there are just rocks there – and my grandmother Joan is leaning against them with a big smile and David is looking very limber.
I have to say I don't like rocks. All I'm saying is that I do not like the way the rocks go out.

So that was it. Your existential seaside crisis. Do you think it still informs which seasides you like today?

I don't think there are any others where those sort of rocks pile down in that absolutely… except near Polzeath actually. Now I realise.

Did anything happen? Like Mrs Moore in the Marabar Caves.

No. It didn't, it didn't. I think it's more to do with. What is it to do with? I mean there is something. I'm not denying it.

I'm looking at it. There's a geology there…

I mean, in terms of your looking at seasides. This is very silly. It's more complex. It's not prom sand. This is… much more…

Raw nature.

Really really hard nature.

And I'm just remembering… I think it was 2012. Easter 2012. And one of the last times I saw my dad. We met you at Mortehoe – Woolacombe way. We found a cove and Dad looked after my stuff and I swam out and – I did the Dart 10k that year so I was fit – and it was hard work and you thought I shouldn't do it, and some woman asked if I was training for the Olympics.

I remember that now –

It was a very sharp, steep cliff and a small cove.

That's right. I'm not saying anything profound, but you can look at a view. I can go to any bit of Cornwall and enjoy a view and then there's all these bits of – you know Margate, Westgate, and even Perranporth gives you the beautiful sand. But Woolacombe. Woolacombe is this weird….

* * *

Fascinating but we'll move on. So round the bay to Minehead. There's a photograph of you and your brother and other children on what looks like a slipway in a slate shingle beach ahead of a sand beach.

We were fascinated by these bleached bones as we walked around, we couldn't ignore the number of animal bones—we hoped they were animal bones that were there—and we began collecting them and without any scholastic knowledge or any help, we were trying to put together skeletons—you know—was this a squirrel, was this a rabbit, was this a vole… and we began to collect them.

Judging by the size of the bones, it's more sheep, lambs… People…

Hahaha.

…mammoths.

We became obsessed by the bones. Anyway, if you're a child on a seaside holiday, this is actually a KEY THING… your parents want you to go and do things: fool about in the pools, dig about in the sand, if you're lucky your parents will come along and help you build a castle, make a boat, make a car, but god if they can get you to…

Categorize some bones…

…they can sit on the deckchairs and not bother about you for hours and hours and hours.

That photograph is 1956 and the Lynmouth floods were 1952, so it was only four years after one of the biggest flash flood events which killed 34 people, one of whom was never officially identified, a young woman. And you said that you thought they were bones from the flood – and that must have been something your father had said, which makes utter sense because wherever you are on that coast… Decomposition in the saline environment… You were taxonomizing…

Absolutely. One of my summer holidays, I spent putting together all these skeletons.

And the year before, in Minehead. You're on a stony beach.

I'm so surprised that it's so stony. My father taught me to swim here. So we went '55 and '56, and ten years later, '63… We used to come from the Sandcombe Hotel and swim before breakfast every morning… It was a very small hotel, and you were identified as people who'd chosen this hotel to spend this terribly precious time and so you were in turn cherished. That was terribly special. Building up this relationship, and we made friends there.

So you stayed in this hotel, overlooking the beach, and your father taught you to swim here. In a lido that's no longer there.

Yes, he was very keen—I was learning to swim in the sea and he realised that to get the confidence and ability and technique we needed the swimming pool so we would go there for a part of each day and he was very systematic and I learnt to swim. And in the evenings we would go and watch swimming galas.

And he taught you.

Yes, and in those days you didn't have armbands – I had waterwings for a while. They were a one-piece thing – almost like a bra with two balloons behind you – so the strap was in front of your breast but behind were two rubber… not balloons… but you know, they would keep you – buoyant. So… I learnt to swim in Minehead (Fig. 4.4).

* * *

August 19, 1955. Joan writes: "Angela's 8th birthday. Got all ready to leave by 3 but Eric busy and we actually got away at 4.30. Birthday tea at Nantwich. RAIN. David sick. Chaos. Got to Gloucester about 10. All exhausted. Rather noisy place".

Fig. 4.4 Eric and Angela with waterwings at Minehead, 1954; taxonomizing, Minehead, 1956 (*Source* © Sefryn Penrose)

August 20, 1955. "Left Gloucester about 9.50. Traffic hold-up at Bridge-water. Eventually arrived at 3.45. Sandcombe Hotel. Seems very nice. Not so impressed with beach so far but children got in it and play and paddle".

In a few days, they were enjoying sandcastle competitions with fellow guests, bathing, and Angela and Eric putting in the hours *swimming* at the lido. Olympic-sized, beautifully provisioned with stands and cafes, diving boards, the lido was built in 1936 by local landowners, but came to a sad stuttering end in 1993 when Butlin's sold the site for development.

I learnt to swim properly in an outdoor pool. I could doggy paddle with the best of them, but my Aunt Joan, from California—another great swimmer—showed me how to do crawl and I remember the moment of understanding. My struggle against the family breaststroke line was over. I refined that crawl in the outdoor cold-water army pool in the village where I grew up. My mother and I swim now in the restored Tinside Lido in Plymouth. Fed by seawater, it blinds in the sun and stands out against the squalls. A treasure from pre-war seaside heights, all the more fabulous for its position along the Hoe, provisioned once with mixed and segregated sea pools, diving platforms in the sea, terraces and bathing huts. Much of this is decrepit and destroyed now, but the swimmers and sunbathers still come. A Saturday morning group of (mostly) women accept all-comers all year round.

* * *

So when I asked you earlier about where you stayed, all your meals were taken care of and you said that was the factor.
Yes, Eric worked very hard all through those years, he really did and it was understood that Joan was part of all that and the reward was she had a holiday where she never would cook anything—not like us with self-catering. He would give her a proper holiday where she would have everything, she would be waited on hand and foot and everything was laid on. Certainly breakfast and an evening meal and if you wanted a midday meal, they would provide sandwiches but, certainly, Joan would have a proper holiday. She loved these holidays so she would plan it for weeks on end. She'd choose a space where she would pile up—everybody had their own little pile—clean pants, clean vests, clean T-shirts or whatever we wore. I don't think we wore T-shirts. So she would buy certain new clothes—we would always get a few new summer clothes. Sun dresses. We were all assuming the sun would fucking shine.
Exhibit A. A photograph from 1955, Linton, Lynmouth, and there's Joan in a lovely summer dress in a really wonderful quite modern pattern, and my mother who would have been eight is wearing a round neck t-shirt and a pair of shorts and David is wearing a stripy t-shirt and a woman who doesn't want to be in the picture and we don't think is part of our gang is wearing a boobtube.
God I am wearing a T-shirt.
That is a hot day, but it is true to say in other days you don't look hot.
But can I just tell you my mother has special earrings on. Special handbag.
It looks like she's had her hair done.
You know, she is on holiday and she can wear stuff. She's never going to wear that any other time of the year and so she's saving up, she's packing, and then as the time came closer, we wore the weirdest clothes because everything we had that was decent was being washed and packed and piled and folded to go in this suitcase to go on holiday so God knows what we wore.

* * *

Joan writes: *August 18th, 1955,* "I went to Cheadle and got Angela a very nice pair of green shorts. Eric worked until midnight." And indeed, she gets her hair done in Minehead, in Dartmouth too another year before going down to the Regal cinema to see Mitzi Gaynor in *There's No Business Like Show Business* – "in Glorious Technicolor" records Joan, "Cinemascope too". "Rather enjoyed it".

George Eliot's visit to these Bristol Channel coasts delighted by another kind of technicolour:

> *I felt delightfully at liberty and determined to pay some attention to sea-weeds which I had never seen in such beauty as at Ilfracombe … The* Corallina Officinalis *was then in its greatest perfection, and with its purple pink fronds throw into relief the dark olive fronds of the Laminariae on one side and the vivid green of the Ulva and Enteromorpha on the other.*

There are the white cilia, the yellow-brown whip of the Mesogloia. "These tide pools made me quite in love with sea-weeds".[24]

* * *

Weston-Super-Mare, no, you never went there.
But we stopped there once when you were little.
And I was disappointed by the mud. But we swam anyway and went all the way home filthy.

* * *

And again, paddling miles into the silky Weston mud the day that Harry and Meghan married (we watched the wedding on the pier with watery Pimms) I met a man and two children going back shorewards. On my way back, I reassure a panicking woman that I had definitely seen them heading in. She knows the tricky mudflats here—which is more than I do, she notes from the mud caked up to my knees. We find her family happily digging in the fine beach sand. She weeps into her partner's shoulder. She'd had a bad experience once in the mud, she'd told me.

* * *

And you never did the south of Wales and we come right up, Aberystwyth? Anglesey?
You missed out Barmouth.
There's an amazing photograph of you aged twelve looking like you're posing for a magazine and Eric has written "my baby aged twelve!". You look about eighteen! You're in an emerald green swimsuit.

* * *

All kinds of women gave the coast. A litany of widows and spinsters bestowed their lands on the National Trust—*in perpetuity*—from Mrs Fanny Talbot who gave her view above Barmouth in 1895; to Agatha Christie's daughter, Rosalind, who gave Greenway in 2000. Rosalie Chichester gave her house and most of her lands at Mortehoe and Woolacombe (aside development in the town) before her death in 1948; Ida Sebag-Montefiore (with an offer of help from Nancy Astor) at Wembury gave her estate to protect it from development; Mary Bonham-Christie gave Brownsea, albeit in lieu of death duties, and through the fight of Helen Brotherton, founder of the Dorset Wildlife Trust. And a litany of widows and spinsters continue to honour the shore and the sea, in perpetuity.

* * *

This picture is of loads of girls changing and then there's another picture with half of you – more than half of you – in bathing caps as you call them – rubberised skull caps.

Criccieth. I loved it because it was my first Guide camp. I've never had so much fun in my life. Most of those people—that was the only time they went to the seaside. We sang Living Doll. Cliff Richard. It was unbelievable, so innocent. And there were boy scouts in the next field.

You could all swim? You're 11 years old.

I could swim. I'm sure most of those people could swim actually.

They all look like great swimmers. In this picture they all do look like they're about to swim the Channel. There're maybe two girls who look like they don't wanna do it and everybody else looks quite enthusiastic. I'm just trying to find you in this picture…

I probably took it!

What was your camera?

A Kodak Box. My father gave me this camera.

Were you the only one with a camera?

Probably.

My grandfather was a photography enthusiast and he bought you a camera.

Look look look – about a third along, there's a woman with dark hair, without a hat. That is Elsie Worthington who is responsible for that.

She has swimmer's shoulders. She has muscles that swimmers have and nobody else has.

Well. She was our Guide Captain so she did this you see. She took us down to the sea and we swam.

When I asked you first about Morecambe. Had you been to Morecambe on holiday, because it is in the North West which is where you grew up... You said you'd been there with the Guides.

With Sunday School! What may not be known universally is that if you go to Sunday School you're allowed to take Ascension Day off, which is the Thursday before Whit Sunday—Pentecost—so these Sunday Schools took all these children off to wonderful places like New Brighton, St Annes, Southport, Blackpool, Morecambe.

And you said that Sunday School took you to the most fucking awful places.

Well, one of the problems—I don't know how people deal with Morecambe—but the sea goes out. So for those of us who want to have a bit of a swim in the waves it's spoilt so you have to go on the rollercoaster. Or buy rock. Candyfloss! So my recollection of that bit of the coast is that it's terribly sad, but my uncle Norman took us in a Standard 8 all the way from Cheshire to Morecambe. And it was absolutely exciting. We stood up in the back seat—oh! no such thing as seatbelts—bottles of Tizer to our teeth so if we'd crashed into anything that would have been.... Anyway, we got there and my uncle said "oh! WHERE'S THE SEA?", but we plunged forward and we had a great time anyway...

In the mud?

Well you know, the sand was very strange.

* * *

Can we end where we started? With the dream of the poetess. Maria Abdy of Upper Marine Terrace, Margate, may she speak for the women who bathe (Fig. 4.5).

Oh! lead me not o'er banks of flowers,

In Summer's garb array'd,
To gaze on fair and fragrant bowers,

And groves of clustering shade;
Proudly I claim my birth and lot,
'Mid scenes more wild and free,
Your woods, your glades, delight me not —

My home is by the sea.

Fig. 4.5 Jago, Angela and Sefryn, Burnham Overstrand, 1982; Perran and Sefryn, Cromer, 1983 (*Source* © Sefryn Penrose)

Notes

1. Maria Abdy, *Poetry,* Second Series (London: Robins & Sons, 1838), 1.
2. Abdy, *Poetry*, 122.
3. Keith Parry, *The Resorts of the Lancashire Coast* (Newton Abbot: David & Charles, 1983), 92.
4. William Blake, *The Complete Poetry and Prose of William Blake* (New York: Anchor Books, 1988), 711.
5. Blake, 723.
6. Blake, 139.
7. Jonathan Roberts, "William Blake's Visionary Landscape Near Felpham", *Blake, An Illustrated Quarterly*, 47, no. 2 (Fall 2013), n.p.
8. Ann Cole, "The Meaning of the Old English Place-Name Element Ōra", *Journal of the English Place Name Society* 21 (1989): 15–22 (18).
9. Violet Hunt, *The Flurried Years* (London: Hurst & Blackett, 1926), 236.
10. B. S. Capp, *When Gossips Meet: Women, Family, and Neighbourhood in Early Modern England* (Oxford: Oxford University Press, 2004), 359.
11. Michael C. Questier, "Conformity, Catholicism and the Law", in *Conformity and Orthodoxy in the English Church, c. 1560–1660*, ed. Peter Lake and Michael C. Questier (Woodbridge: Boydell & Brewer, 2000), 237–261 (242).
12. Hunt, *The Flurried Years*, 242.

13. AA (Automobile Association), *Illustrated Guide to Britain's Coast* (London: AA, 1985), 89.

14. Caroline Sedgewick, *Letters from Abroad to Kindred at Home* (London: Edward Moxon, 1841), 12.

15. Countess of Cardigan & Lancaster, *My Recollections* (New York: John Lane, 1909), 25.

16. Tammy M. Proctor, *Scouting for Girls: A Century of Girl Guides and Girl Scouts* (Oxford: Praeger, 2009), 20.

17. Jane Austen, *Persuasion* (New York: The Novel Library, Pantheon Books, n.d.), 109.

18. Agatha Christie, *An Autobiography* (New York: Dodd, Mead & Company, 1977), 130.

19. Christie, 133.

20. All dictionary references are to the OED, Third Edition, 2016.

21. Norman Longmate, *Defending the Island: From Caesar to the Armada* (London: Pimlico, 2011).

22. Charles Kingsley, "Sir Walter Raleigh and His Time", *North British Review* (1855), 252.

23. John Buchan, *Sir Walter Raleigh* (London: Thomas Nelson), 1911.

24. George Eliot, *The Journals of George Eliot*, ed. Margaret Harris and Judith Johnston (Cambridge: Cambridge University Press, 1998), 264.

5

Sands, Good Sands, Excellent Sands: Writing and Ranking the British Coastline in the Middle of the Twentieth Century

Tim Cole

Sands,
Good Sands,
Excellent Sands.

Fine Sands,
Lovely Sands,
Splendid Sands.

Magnificent Sands,
Suave Sands,
Singing Sands.

Extensive Sands,
Expansive Sands,
Teeming Sands.

These sandy descriptions are found scattered among the gazetteer entries at the back of a series of mid-century guidebooks to Britain.[1] As they directed tourists to the British seaside, travel writers offered their take on the quality of the sand that they would find. In part, they replicated a geography of

T. Cole (✉)
University of Bristol, Bristol, UK

© The Author(s) 2020
J. Carruthers and N. Dakkak (eds.), *Sandscapes*,
https://doi.org/10.1007/978-3-030-44780-9_5

coastal Britain made up of the material stuff found on its beaches: sand, shingle, stone. But gazetteer writers also created an imaginative geography of differing types and qualities of sands, that ranged from those deserving no adjective, through "good" and "excellent" and—in the case of H. G. Stokes whose words I cite above—far, far beyond. But just as the sands themselves were shifting, so were the ways that they were described in the tumultuous decades in the middle of the twentieth century. These decades saw a shift in how visitors accessed—and thought about—the coast, as the age of rail was overtaken by the automobile age. This shift in transport technologies was entwined with a move from thinking about the sea-line predominantly as a scattered geography of sandscapes to a more continuous linear geography of coastal landscapes. As I explore in this essay, both sandscape and coastal landscape were seen as ripe for classification and ranking, with the very best offered up to be visited or conserved.

The ranking of place emerged as a staple within guidebooks published to service the new business of mass tourism.[2] Most famously, the Baedeker guides adopted a star system—that was then subsequently widely copied—to distinguish tourist sites from each other. This simple rating system identified a handful of "must see" sites that offered the time-pressed tourist lacking local knowledge with a ready-made itinerary that included the best of the best. The creation of hierarchies within guidebooks became a norm, although not everyone adopted a system of stars. Other methods were deployed to distinguish between good and best. In the case of Stokes, whose varying sandscapes preface this essay, this was done by careful use of a wide range of adjectives.

Stokes' sandscapes are found at the back of a series of guidebooks that were published in 1951 by the Festival of Britain Office.[3] Breaking the country down into regional areas, these self-proclaimed "new" guidebooks were intended to carry the Festival message and bring it to life as visitors made their way *About Britain*. In the aftermath of war, the Festival painted a picture of Britain rising from the ashes and very much open for business. This was the central story found at the main exhibition on the South Bank in London, but it was also one that visitors were encouraged to discover for themselves as they drove *About Britain*. Published by Collins, with the financial support of the Brewers' Association, each of the thirteen volumes contained a lengthy "verbal portrait" that delved deep into the geology, archaeology and history of the region through the penmanship of a local "expert". At the back, six-day-long driving tours of each area were recommended, along with accompanying strip maps, captions and thumbnail sketches. As each guidebook explained, these had "been prepared with the aid of local experts" and "should give you the essence of town and countryside with the least expenditure of time".[4]

For those with more than a week on their hands, a gazetteer detailing sites of interest in the major towns and villages across the region completed each volume. While a variety of local "experts" were appointed to write the verbal portrait, or to check the tours on the ground, the gazetteer was left—two volumes aside—to one man: H. G. Stokes.[5] Stokes had been brought in as Editor for Content when the Brewers' Society offered their sponsorship of the series in November 1949.[6] It was obviously assumed that gazetteer writing did not require the same level of place-based expertise that each regional "verbal portrait" demanded. Instead, what was seen to be important in accomplishing this more technical exercise was uniformity of approach.

As he set about writing these gazetteer entries, Stokes worked his way imaginatively around Britain's shoreline and evoked the kind of sands that visitors would discover along their journey. In part he classified the British coast as a shoreline of differing geological materiality. Beaches varied. Sand was not found everywhere. Rather, the British coast was made up of a patchwork of "sandy", "sand-shingle" and "shingle" beaches depending where you were. But Stokes offered more than simply a geological guide to the British coast. Guidebooks do not simply describe. They also make value judgements. And so Stokes created a hierarchy of British resorts from those places that possessed simply "sands", through to those with "good sands", "excellent sands", "fine sands" and beyond.

Here his approach fitted with the language and method adopted by other mid-century gazetteers found in the glove box of the middle-class motorist. The most important of these—the Automobile Association's *Road Book*—contained a gazetteer which included many of the same phrases that Stokes deployed when referring to the British coastline. Like Stokes, the AA gazetteer signalled the geological make-up of beaches, pointing out which places were "sandy", which were of more mixed fare and were "pebble and sandy" or "sand, rock and shingle" and which were simply the home of a "pebble beach".[7] Also like Stokes, the AA *Road Book* sought to alert its readers to markers of the quality of sand found at each of those beaches that were "sandy". Some beaches merely had "sands", while others had "good sands",[8] "excellent sands"[9] and "fine sands".[10] Just as gazetteer writers signalled which of the many thousands of parish churches across the country were worth stopping the car so that the motorist could explore their architectural interior, so some stretches of sandy beach were highlighted as places to get out of the car and get the sand—and particularly good sand at that—literally between your toes.

However, there was not complete agreement between gazetteer writers over which beach—nor church for that matter—was the finest. In a very small number of cases, these lists were unanimous in their rankings. Stokes and the AA *Road Book* were in complete agreement when it came to five beaches. Both described Abersoch, Barmouth, Colwyn Bay and Morfa Nevin as having "good sands", while Clacton-on-Sea possessed "excellent sands". However, this is where their agreement ended. In the majority of cases, they differed in their choice of adjectives and created rather different imaginative geographies of Britain's sands. This is not surprising. This was not the scientific measurement of water quality deployed in more recent European hierarchies of sea-water quality and the awarding of Blue Flag status. The description of sands by gazetteer writers was a more subjective and creative process.

Stokes—and in the case of the first volume to "Wessex", series-editor Geoffrey Grigson—constructed in the pages of the *About Britain* guides an idiosyncratic hierarchy of the British coastline which stretched from those places with mere "sands"—[11] Bude, Carbis Bay by St Ives, Bournemouth, Margate, Saundersfoot, Southerndown, Llandudno, Prestatyn, Mablethorpe, Formby, Robin Hood's Bay, Staithes, Arbroath, Troon and on Islay—to those many more places with "good sands". These were found aplenty in Stokes' Britain, as he worked his way from Devon and Cornwall, first east along the south coast and up into East Anglia, then west into Wales before finally heading northwards up the Lancashire, Yorkshire and Northumberland coasts and into Scotland.

Stokes' Britain was an island blessed with "good sands" at Exmouth, Teignmouth, Studland, Cromer, Gorleston, Lowestoft, Mundesley, Overstrand, Walton-on-the-Naze, Aberystwyth, Llanstephan, Aberdovey, Abersoch, Barmouth, Borth, Colwyn Bay, Holyhead, Hoylake, Nevin, Portmadoc, Pwllheli, Rhyl, Bridlington, Cleveleys, Hornsea, Withernsea, Seascale, Whitley Bay, Crail, Banff, Burghead, Durness, Fortrose, Helmsdale and Nairn.[12] Dotted around this coastline made up of "good sands", Stokes picked out a few places that demanded a range of greater superlatives. There were "excellent sands" at Clacton-on-Sea and Southport.[13] "Fine sands" were found at Newquay, Caister, Aberffraw and Newbiggin by the Sea.[14] Glenluce and Gruinard Bay had "lovely sands".[15] Weston-super-Mare possessed "magnificent sands".[16] Drawn to alliteration, Stokes declared that Skegness and Rhossili Bay offered "splendid sands" and Great Yarmouth had "suave sands".[17]

When it came to describing which sands were worth stopping to see and feel, Stokes outdid other gazetteer writers with the range of adjectives he deployed. Alongside the triplet of "good sands", "excellent sands" and "fine sands" that you find in the industry standard of the Automobile Association guide, Stokes added his own idiosyncratic set of adjectives as he sought to differentiate one sandy beach from another and signal that all sand was not the same. His attempt to capture the atmosphere of the coast brought forth a flow of adjectives. Seaton Carew, Redcar and Montrose had "extensive sands".[18] "Cockle Beach" at Barra had "a wonderful expanse of sand and shell".[19] Iona and Morar were both famed for their "white sands", which, on Iona "render the colours of the sea more intense than elsewhere".[20] At Lunan Bay, the visitor would find "singing sands", and at Perranporth a "two-mile stretch of hard sands".[21] At Ramsgate, they would encounter, "teeming sands".[22] In his use of a wide range of superlatives, Stokes not only drew away from the norms of other gazetteers, but also it seems the guidance drawn up by the Festival of Britain Office.

Plans for this series of guidebooks evolved over 1949, with the final shape of the gazetteer index finally being decided at the third meeting of the Guide Book Editorial Committee in January 1950. At this meeting it was reported that the publishers had encountered, "some difficulty [...] in determining the sort of information that should be included in the Gazetteer index", and so the Editorial Committee was quick to offer its advice. They determined, "that information of factual nature was necessary and that in all cases the 'atmosphere' of the place should be indicated as comprehensively as space would permit which would undoubtedly mean in general, very briefly". Word count was extremely limited in these volumes that would each contain only ninety-two printed pages. Given these constraints, only "three or four" of the "most important towns" in each region were to be described in "say 200 words". There then followed a hierarchy of descending importance, with "towns and places of medium importance" given "say 100/150 words", "small towns and places of lesser importance" to be described in "50/100 words" and relatively unimportant "villages" in "say 25/50 words". With words at such a premium, and with an eye to the importance of factuality, adjectives—"such words as – beautiful, charming, etc."—were explicitly to "be avoided".

To flesh all this out, the Festival Office created "a draft specimen gazetteer entry for Norwich containing about 200 words" which all "agreed [...] was exactly the sort of thing wanted and that this, together with the general scheme outline above should be passed to the Gazetteer writer as a guide".[23] Sadly this draught gazetteer entry for Norfolk does not survive in the archive. But we can, perhaps, assume that Stokes drew heavily on it—and may have simply copied it—when drawing together the gazetteer for the fourth volume in the series on East Anglia. In this, Norwich, Norfolk, is tersely described in less than 200 words as:

A centre for agricultural marketing and industries, such as agricultural equipment, mustard, boots and shoes, silk-weaving (established by the Huguenots in 17th century), etc. The Norman Castle was used as a county jail up to 1894, since when noted as museum, including art gallery rich in paintings of the Norwich School – John Crome, Cotman, Thurtle, Starke and Ladbrooke. Unrivalled view of city from Keep – "a church for every week; a pub for every day." Note striped awnings in the market square. The Cathedral is the only English cathedral without a crypt (due to location on marshy ground). Mainly Norman though spire much later. Surrounding buildings damaged in blitz.

The churches, the Grammar School, St Helen's Hospital with its ancient Swan Pit; remains of monastic institutions, the leper house (now part of public library); the Curat House and many other places of historical association combined with present use are some of the interesting buildings. Many pleasant walks; boating on the rivers (Lenwade and Yare). Maddermarket Theatre.[24]

With few adjectives—"interesting buildings" and "pleasant walks" aside—the entry chimed with the hopes articulated for these guides that, "the main thing is to show clearly what goes on in an area, so that a man interested in painting won't go to Glasgow and miss the Kelvinside Art Gallery, etc.".[25] There was little chance of anyone visiting Norwich, armed with their *About Britain* guidebook, missing much of importance, and certainly not the collection of paintings by Norwich School artists in the city art gallery.

However, the aim to offer succinct advice in these "new" guidebooks to anyone interested in knowing "what goes on in an area" was not entirely successful. In a damning review of the *About Britain* guide to the *East Midlands and the Peak*, Stoke's gazetteer was dismissed as "slipshod" with "three major blunders in an entry on Leicester that is not a hundred words long".[26] This was, no doubt, a result of deploying a single author for the gazetteer entries in eleven of the volumes, the East Midlands included, who lacked local expertise. But while Stokes may have got Leicester wrong, he did seem to know a thing or two about sand. A few years earlier, he had published a history of *The English Seaside* that traced the emergence of seaside resorts in the mid-eighteenth century with the rising popularity of sea-water bathing and their subsequent flowering in the nineteenth century.[27] As Stokes explained in this popular history, attention shifted in the nineteenth century from sea water, via the sandy beach, to the full-blown resort-town—in the case of post-railway-age Blackpool—complete with "piers, parades, ballrooms, the Big Wheel, the Tower, houses, hotels, boarding houses…".[28]

Although his focus in the gazetteer entries was on ranking the various qualities of the sands found on British beaches, Stokes was well aware that visitors to the post-railway-age British coast would find more than simply sand of

varying quality. Turning his attention to what lay the other side of the promenade, Stokes also used his gazetteer entries to signal gradations in the quality
of the seaside architectural vernacular, not simply the sand. He was particularly critical of the architecture found at those many resorts developed in the
second half of the nineteenth century. "One of the mysteries of the Victorian
age", Stokes declared in the pages of *The English Seaside*, "is the sudden disappearance of a national artistic sense", as the architectural influences of "Wren,
Vanbrugh and the Regency period" were replaced by buildings "decked out
with bits of meretricious ornament" and "*cottages ornées* sprung up all around
the coast".[29]

The result of close to a century of resort-town building was that while
some places might boast wonderful sands, their architecture left much to be
desired. Case in point for Stokes was Newquay in Cornwall that had "fine
sands", but "its late development" had left it with "hardly the most elegant
architecture".[30] Criticism was levelled not only at those resort towns developed during the age of rail, but also at those coastal ribbon developments
established during the age of road. One such place was the Gower peninsula
on the south coast of Wales, where "a splendid stretch of sand" was spoilt by
what Stokes dubbed, "inevitable bungaloid settlements on the coast".[31] By
contrast, those resorts developed far earlier were praised not only for their
sands but also the buildings abutting these. For example, Aberystwyth on the
Welsh coast had been developed several decades earlier than Newquay and
therefore had, not only "good sands", but also "some pretty Early Victorian
buildings and a refreshing absence of spit and polish and chromium".[32]

Like the Festival of Britain as a whole, the guides tended to celebrate the
stream-lined aesthetic of Georgian neo-classicism over the high gothic decora-
tion of late Victoriana or the speculative building of the 1920s. In some cases,
as with series-editor Geoffrey Grigson's withering dismissal of Bournemouth
on the south coast, visitors were advised to look out to sea and pretend that
the town was not there at all. Adopting this strategy, "the view from the cliff
top near the gallery of the long sands, the bathers, the pleasure boats cutting
white arcs in the blue sea, makes one forget much of the less pleasant urban
landscape of the town".[33] This invitation to turn their back on late Victorian
and early twentieth-century architecture and gaze seaward was one that Stokes
offered to visitors to the "progressive" seaside resort of Morecambe. There was
no doubt too much "spit and polish and chromium" at the Midland Hotel
for Stokes' liking, and he urged visitors to ignore the town and look across
the bay to discover the "magnificent views of the Lakeland mountains from
the sandy beach".[34]

While Stokes and Grigson, and the wider Festival of Britain, adopted a specific mid-century-modernist critique of late nineteenth-century Victorian seaside-resort architecture, their criticism of relatively unregulated seaside-bungalow building in the early decades of the twentieth century was much more widely shared. As middle-class motorists accessed previously inaccessible stretches of coastline, their ramshackle holiday homes became a staple of critics of the automobile age and the changes it brought to the British landscape.[35] But the motoring age, and what were seen as the unsightly developments that came in its wake, also resulted in a more profound shift in ways of thinking about the shoreline. While Stokes was busy working his way, imaginatively, around Britain and ranking the quality of its sand, others were also circumnavigating Britain and ranking the quality of its coast. Where saltwater met shore, was not simply seen as sandscape, but increasingly as coastal landscape.

While the Second World War was still being fought, a contemporary of Stokes, J. A. Steers was invited to address the Royal Geographical Society. He spoke to them on 5 June 1944—on the cusp of D-Day—in a building that had recently been repaired following bomb damage.[36] Steers' thoughts were already turning to the post-war world and "the future of the countryside" in general, and the British coastline in particular. In his lecture, Steers presented the results of an extraordinary one-man survey of the British coast. He had walked 2751 miles of the English and Welsh coast over the course of the preceding eighteen months—come "rain or shine"—"with the exception of those south-eastern parts, which for obvious reasons he could not visit".[37] Surveying those stretches of coast that were furthest away from the frontline of the war, Steers attempted a country-wide "assessment of coastal quality", that sought to classify stretches of coast "into classes based on a constant standard".

His one-man mission to describe and rank the coastline bore some resemblance to Stokes' rating of British sandscapes as he took visitors *About Britain*. Rather than simply offering a binary distinction of the British coast into those stretches that were "spoiled" and "unspoiled", Steers sought to offer a gradation of value. However, he wanted to keep things simple. Rejecting an "elaborate classification of coastal quality", he stuck to a simple threefold ranking. Just as mid-century gazetteers distinguished between "good sands", "excellent sands" and "fine sands", Steers offered three categories of coastal beauty—"good", "very good" and "exceptional".[38] Acknowledging that this was a subjective exercise and that no one "could claim an absolute standard", Steers was still confident that his one-man approach offered a consistency that meant that, "it does not matter if the standard is too high or too low: it can be easily adjusted". In short, he argued that whether you agreed with the absolute categories or not, his relative ranking was seen to be of value.

Steers was well aware that his rankings of particular stretches of coastline might be challenged. However, he was fairly confident in his choice of a small number of sections as "exceptional". Thus he assumed that "most people would probably" agree with his ranking of "the Mawddach and Dyfi estuaries, parts of the Pembrokeshire coast, the Hartland district, and the southern extremity of Devon" as "outstanding". It was the more intermediate category—"very good"—where he acknowledged that there was likely to be a "considerable difference of opinion". But these differences in opinion came more quickly than he perhaps ever imagined.

Shortly after sitting down, Steers' ranking was questioned in the discussion that immediately followed his paper. The focus of attention was Morecambe Bay, which Steers had rendered as merely "good". This was challenged by Professor A. G. Tansley who argued that he would, "put the unspoiled parts" of Morecambe Bay, "particularly the north shore, into a higher class than Mr Steers does". While he acknowledged that this coastline was "not spectacular" he leapt to its defence as a coast that was,

> extremely beautiful and varied with salt marshes, sandy patches, outcrops of rock, and little woods here and there. It is still quite remote and in parts access is rather difficult. This coast has great scenic charm of a quiet kind as well as varied ecological interest. It is well worth safeguarding, for there are few stretches of our coast which have these particular merits.[39]

One of the founding figures in the new science of ecology, Tansley approached the coastline of Morecambe Bay somewhat differently to Steers: the former as distinctive ecosystem and the latter as visual spectacle. Adding Stokes into the mix, with his critique of Morecambe's resort architecture, it is clear that there were radically different ways of seeing the British coastline and therefore also in ranking it.

What the exchange between Steers and Tansley points to, is that for Steers, it was the look of the coast that was of primacy. This comes through clearly in his justification of including "the Hartland district" of Devon in South-West Britain within those limited stretches of coastline he deemed "exceptional" or "outstanding". Although he acknowledged that, "Exmoor is beautiful, and the adjacent coast very fine", he was quick to signal that, "it cannot stand comparison sensu stricto with the exposed coast, magnificent cliffs, and intricacy of detail that exist between Hartland Point and Morwenstow". There was no reference to the "varied ecological interest" that Tansley saw as so important, but neither was there any reference to the sands that Stokes cared so much about. This "exceptional" stretch of British coast around Hartland Point was not made up of beaches to be played on, but of cliff-tops to be walked along.

While Steers shared Stokes' assumption that the coastline could be classified into relative categories, he saw the British coast very differently. It was not first and foremost a sandscape, but a cliff-scape that offered up "long lines of unspoiled coast" as a "paradise of the walker and naturalist".[40] Stokes was interested in the quality of "sands" in seaside resorts, but Steers was more interested in the quality of "natural scenery" along the whole stretch of coast. Steers imagined Britain as made up of lengthy stretches of "spoiled" and "unspoiled" coastal scenery (an estimated 2751 miles around England and Wales), Stokes had a more staccato approach that isolated spots of "good" and "excellent" sands found at the centre of resort towns of varying architectural finesse.[41]

This movement in imagining the British coastline from a series of resort nodes to a coastal network signalled, in part, the shift in transport technologies across the nineteenth and into the twentieth century. Instead of accessing the sandy beaches of those distinct seaside resorts that grew up at the end of the rail tracks, the twentieth century saw the opening up of the far larger stretches of the British coastline that lay at the end of the rapidly growing road—and footpath—network. In this context of the greater access the motor car afforded, came calls to safeguard this newly accessible coastline through nation-wide regulation of planning that Steers was clearly a leading advocate of. Mirroring the twin-aims of emerging national parks, this was an attempt to marry the aims of access to, and preservation of, "wild" landscapes in the automobile age.

The idea of preserving "unspoiled" stretches of coastline was crucial to Steers, and this is where sand was a problem. He was quick to acknowledge that "sand or shingle beaches give great beauty and character to our coasts". However, sandy beaches were honeypots that tended to attract the unplanned bungalow and "hut" building that he saw "spoiling" the British coastline. While Stokes was quick to make judgements on the quality of seaside architecture, Steers took this critique one stage further. He did not share Stokes' preference for early nineteenth-century architecture over that of the late nineteenth century and early twentieth century. Instead, any unregulated building—whatever its architectural merits—threatened to "spoil" the British coastline.

Steers and Stokes were united in their criticism of "bungaloid settlements" along the coast. As Steers explained in his lecture, while walking around the coast, "particular notes were made of all shacks, huts, ugly, and misplaced buildings". But his criticism extended much further to include all "poorly planned and sited" structures. As he explained, "good, well-built, and often artistic houses may be quite as offensive as meaner dwellings, especially if they are on open cliffs or if the individual houses or groups are poorly planned and sited", adding that, "a tour round parts of Anglesey and Cornwall illustrates this point very well".[42] For Steers, it was only those stretches of coast that remained "unspoiled"—whether "cliffs, dunes, salt marsh, estuary"—that "should be rated basically as good natural scenery".[43] And his call was for the post-war state to regulate building around the coastal belt through a national authority rather than the more piecemeal approach of local authorities' oversight over their own domain that he saw as having resulted in so much coastal spoliation.[44]

This signals a critical difference between Stokes and Steers as they worked their way around the British coastline in acts of classification. They had very different audiences in mind. Steers' ranking was not found in the pages of mass-produced guidebooks, but was marked up on a map included alongside the text of his lecture in *The Geographical Journal*. While Stokes sought—in the manner of the guidebook writer—to direct the visitor to this, rather than that, beach, Steers aimed to influence policy makers' attitudes towards those coastal stretches that he saw to be of high scenic value.

At least one such policy maker was in the audience when Steers read his paper in 1944. The Minister of Town and Country Planning—William Morrison—was not only present, but responded by thanking Steers for his survey and "all the expert help he has given us over the last eighteen months", and signalled an intent to develop "a partnership of central and local planning authorities", to safeguard and open up "the unparalleled beauty of our coastline".[45] Steers' survey fitted within a broader set of concerns to conserve and provide access to the countryside that lay at the heart of planning for postwar National Parks in England and Wales, although these—Pembrokeshire apart—were to be found in the uplands not the coastlands.

What Stokes' gazetteer entries, Steers' one-man mission to walk and rank the coastline, and Tansley's challenge over the precise status of Morecambe Bay point to are a moment in the mid-twentieth century when a number of rather different discourses of the British coast co-existed. As I have suggested, one way of seeing this is as a shift away from focusing on sand—and the isolated beach—to the coast as a networked whole. This meant that more of the coast was opened up as possessing value. As Steers' praise of the cliffs around Hartland Point in Devon show, this included rocky stretches of coast where sand was in limited supply or was largely inaccessible. But it also included those more muddy stretches of marshland, where estuarial mud and salt marshes—like those at Morecambe Bay—were included in the mapping of valuable coastal scenery.[46] It is clear that in the mid-twentieth century the tide was shifting. While the quality of the sands remained of importance to gazetteer writers and their imagined audiences, other voices were praising the rocky cliffs and muddy margins and their scenic and ecological value. This was a moment when ways of seeing the shoreline were changing, and coastal Britain was seen as made up of more than simply spots of good and excellent sands.

Notes

1. Geoffrey Grigson (ed.), *About Britain* (London: Collins, 1951) (hereafter *About Britain*).
2. Rudy Koshar, "'What Ought to Be Seen': Tourist Guidebooks and National Identities in Modern Germany", *Journal of Contemporary History* 33, no. 3 (1998): 323–340.
3. *About Britain*.
4. See "The Tours" introduction in each volume of *About Britain*.
5. The gazetteer for *About Britain* volume 2, Wessex, was written by General Editor of the series, and author of the Verbal Portrait of the Wessex guidebook, Geoffrey Grigson. The gazetteer for volume 13, Northern Ireland was written by Hugh Shearman. Shearman did not pay attention to sands in the way that Stokes did, although he did point out a few towns which had a "good strand", "fine strand" or "beautiful strand": see *About Britain* no. 13, 84, 89, 92.
6. The National Archives, Kew (hereafter TNA) WORKS 25/57/A5/Q2 (4 November, 1949). Stokes was the author of, among other books, the 1941 Penguin Special, *Signalling and Map Reading for the Home Guard* and *English Place Names* (London: B.T. Batsford, 1948).
7. *AA Road Book of England and Wales* (London: The Automobile Association, 1950), 138–140.
8. "Abersoch", "Barmouth", "Burnham-on-Sea", "Camber", "Carlyon Bay", "Colywn Bay", "Cullercoats", "Deganwy", "Felpham", "Goodrington", "Gorleston-on-Sea", "Hayle", "Hayling Island", "Hunstanton", "Kingsgate", "Morfa Nevin", "Mundesley-on-Sea", "Paignton" "Prestatyn", "Rock", "Sandbanks", "Seahouses", "Seaview", "South Shields", "Thurlestone", "Wallasey", "Wells-next-the-Sea", "Westgate-on-Sea", "Weston-super-Mare", "Weymouth", "Winterton-on-Sea" in *AA Road Book of England and Wales* (London: The Automobile Association, 1950), 139, 148, 174, 177, 181, 194, 199, 201, 217, 224, 233, 243, 251, 289, 290, 308, 318, 325, 332, 334, 345, 360, 367, 373, 374, 375, 376, 383.

9. "Blackpool", "Blyth", "Bognor Regis", "Boscombe", "Bournemouth", "Broad-stairs", "Bude", "Clacton-on-Sea", "Cleethorpes", "Climping", "Exmouth", "Frinton-on-Sea", "Hornsea", "Hoylake", "Ingoldmells", "Mudeford", "Newbiggin-by-the-Sea", "Newquay", "Padstow", "Par", "Penmaenmawr", "Perranporth", "Polzeath", "Port Erin", "Port Talbot", "Ramsey", "Ramsgate", "Rhyl", "Ryde", "Porthminster, St Ives", "Sandown", "Sansend", "Selsey", "Sennen", "Shanklin", "Skegness", "Swanage", "Swansea", "Tenby", "Totland Bay", "Walton-on-Naze", "Waxham," "Whitby", "Whitley Bay", "Widemouth Bay", "Withernsea" in *AA Road Book of England and Wales* (London: The Automobile Association, 1950), 159, 160, 161, 162, 169, 173, 190, 191, 216, 221, 241, 242, 245, 290, 292, 295, 308, 310, 311, 315, 316, 320, 323, 328, 330, 333, 335, 337, 340, 355, 358, 362, 369, 372, 377, 378, 379, 384.

10. "Oxwich", "Pentraeth", "Prah Sands", "Saunton Sands", "Seascale," in *AA Road Book of England and Wales* (London: The Automobile Association, 1950), 307, 310, 318, 333, 334.

11. *About Britain* vol. 1: 82, 88; *About Britain* vol. 2: 81; *About Britain* vol. 3: 88; *About Britain* vol. 6: 91; *About Britain* vol. 7: 86, 89; *About Britain* vol. 8: 88; *About Britain* vol. 9: 82, 88; *About Britain* vol. 10: 91; *About Britain* vol. 11: 83, 91; *About Britain* vol. 12: 88.

12. *About Britain* vol. 1: 84, 89; *About Britain* vol. 2: 89; *About Britain* vol. 4: 80, 82, 85, 86, 91; *About Britain* vol. 6: 82, 87; *About Britain* vol. 7: 80, 81, 82, 83, 85, 86, 88, 89; *About Britain* vol. 9: 81, 83; *About Britain* vol. 10: 90, 92; *About Britain* vol. 11: 84; *About Britain* vol. 12: 83, 84, 86, 87, 89.

13. *About Britain* vol. 4: 80; *About Britain* vol. 9: 89.

14. *About Britain* vol. 1: 87; *About Britain* vol. 4: 79; *About Britain* vol. 7: 80; *About Britain* vol. 10: 89.

15. *About Britain* vol. 11: 87; *About Britain* vol. 12: 83.

16. *About Britain* vol. 1: 90.

17. *About Britain* vol. 8: 90; *About Britain* vol. 6: 85; *About Britain* vol. 4: 92.

18. *About Britain* vol. 10: 87, 90; *About Britain* vol. 11: 89.

19. *About Britain* vol. 12: 83.

20. *About Britain* vol. 12: 87–89.

21. *About Britain* vol. 11: 89; *About Britain* vol. 1: 87.

22. *About Britain* vol. 3: 89.

23. TNA, WORKS 25/57/A5/Q2, Minutes of the Third meeting of the Guide Book Editorial Committee held at Messrs. Holden at 94 Mount Street, W1 on Friday 27 January, 1950.

24. *About Britain* vol. 4: 85–86.

25. TNA, WORKS 25/57/A5/Q2, Guide Book Editorial Committee, 26/9/49, "Notes on Financial and Other Implications of 'Happy Travellers' Guides for the Festival of Britain" (26 September, 1949).

26. Jack Simmons, "Review of *East Midlands and the Peak* by W.G. Hoskins", *Leicestershire Archaeological Society*, no. XXVII (1951): 108.

27. H. G. Stokes, *The English Seaside* (London: Sylvan Press, 1947).

28. Stokes, *The English Seaside*, 55.
29. Stokes, *The English Seaside*, 70.
30. *About Britain* vol. 1: 87.
31. *About Britain* vol. 6: 85.
32. *About Britain* vol. 6: 82.
33. *About Britain* vol. 2: 81–82.
34. *About Britain* vol. 9: 87.
35. C. E. M. Joad, *A Charter for Ramblers* (London: Hutchinson & Co., n.d.), 121–127.
36. The Right Hon. Sir George Clerk, "Address at the Annual General Meeting of the Society, held on 19 June 1944, by the President", *The Geographical Journal* CVI, nos. 1–2 (July–August 1944), 1.
37. Comments of the Minister of Town and Country Planning, Mr W. S. Morrison, *The Geographical Journal* CIV, nos. 1–2 (July–August 1944), 19.
38. J. A. Steers, "Coastal Preservation and Planning", *The Geographical Journal* CIV, nos. 1–2 (July–August 1944), 7.
39. Professor A. G. Tansley, "Coastal Preservation and Planning: Discussion", 22.
40. Steers, "Coastal Preservation", 8.
41. Steers, "Coastal Preservation", 17.
42. Steers, "Coastal Preservation", 8.
43. Steers, "Coastal Preservation", 8.
44. Steers, "Coastal Preservation", 17.
45. Comments of the Minister of Town and Country Planning, Mr W. S. Morrison, *The Geographical Journal* CIV, nos. 1–2 (July–August 1944), 18, 20.
46. Steers, "Coastal Preservation", 9.

6

Queer Sands: Passion and Dynamic Sexualities in the Edwardian Sandscape

Nour Dakkak

"You can't build a house on the sand, and passion's sand. We want bed rock", explains Clive to Maurice, his Cambridge sweetheart and the eponymous protagonist of E. M. Forster's posthumously published novel, when discussing the future of their precarious relationship in early twentieth-century England.[1] Used as a metaphor for love and passion in the novel, sand implies unpredictability and recklessness. Sand is unreliable and untrustworthy, unsuitable as a foundation. Passion—an intense emotion, impulse or overpowering feeling—is likewise linked here with the temporal, worldly and earthly. This chapter considers the ways in which the Edwardian writers E. M. Forster and D. H. Lawrence turn to sand when they explore the instability of love and passion, where sandscapes become spaces to contemplate and write about the uncertainties and unsettledness of male protagonists' sexual identities.

In his critically acclaimed novel, *Sons and Lovers* (1913), Lawrence uses the Lincolnshire coast as a space in which the protagonist, Paul Morel, evaluates his romantic relationships and his sexuality. Written around the same time, Forster's *Maurice* (1971 [1913–1914]) explores similar themes on an unidentified British coast where the fourteen-year-old boy Maurice and his teacher take a walk and have a conversation about sex in a scene that sets the tone for the whole book. Critics are prone to sift the writings of these two notoriously queer writers for biographical treasure. Indeed, what is often most celebrated

N. Dakkak (✉)
Arab Open University, Ardiya, Kuwait

© The Author(s) 2020
J. Carruthers and N. Dakkak (eds.), *Sandscapes*,
https://doi.org/10.1007/978-3-030-44780-9_6

about their writings is how the facets of their hidden-yet-admitted homosexuality appear within their fictional works.[2] While both writers are drawn to the sandscape in ways that allow them to explore themes of transformation and change, the environment's instability suggests that fiction, similarly, offers insubstantial ground for strict biographical unearthings.

In their writings, the seaside is portrayed as a physical and creative space that challenges fixed and rigid notions of human identity. Sociological research on the British seaside has continuously sought to contest static representations of space and asserts instead a more open and inclusive understanding of the seaside as a liminal environment. Daniel Burdsey argues in his book *Race, Place and the Seaside: Postcards from the Edge* (2016) that

> the English seaside has never possessed a solitary purpose or meaning [...] coastal environments are complex, ambiguous, and fluid landscapes. The seaside can be read and experienced in multiple ways in relation to its functions, attributes, and charms, as well as its peculiarities, hazards and repulsions.[3]

Burdsey notes the dynamic social structure of coastal towns, which host complex types of movements (human and non-human) that characterise them as "fluid landscapes". Burdsey also uses the term "coastal liquidity" in his book to describe the various types of formations that take place in coastal environments, and to explain their dynamicity and unpredictability. While using terms such as "fluid" and "liquid" are helpful for understanding the vibrant character of the seaside, we see Lawrence and Forster turning not to the sea, but to the materiality of sandscapes instead. Sand, like water, is dynamic. Its granularity and texture, the result of lengthy and continuous natural and/or industrial processes, offer images that help us to generate a better understanding of how seaside environments are not only socially dynamic but also materially unpredictable.

Beaches, literally and metaphorically, are often understood as temporal spaces and when they do act as homes, it is to various types of social, cultural and environmental changes. Both Lawrence and Forster respond to the instabilities of coastal environments in their writing as characters engage visually and sensually with the movement of elements which leave direct yet uncertain destabilising effects on them. Sandscapes are spaces that are socially and geographically uncategorised. Because they are distanced from traditional social norms and culture, they embrace difference and become spaces of the possible and potentiality.

When writing on the seaside, the two Edwardian writers deliberately remove it from wider social, cultural and economic contexts that characterised and defined human relationships at the beginning of the twentieth century. The shoreline offers a different kind of site in the uninterrupted movement of waves on sand, of the mutability of the sandscape itself in ever-shifting sand grains, and in the novels' contingent and transient occurrences which take place on the shore. The impermanence and morphology of sandscapes may correspond to the arguments central to queer theory, in which human identity, including sexual identity, is multiple, unstable and unconfined. Queer theory undermines notions of essences—that people are singular—and deconstructs binary oppositions in relation to sexual, and other, identities. Although it is built upon gender and sexuality studies, queer theory has been heavily reliant on postmodern concepts which argue that identity is constantly switching among different roles, positions and potentialities. While the term fluidity is frequently used by queer theorists to question stability and unity, Lawrence and Forster notably turn towards the granular qualities of sand when writing about human identity.[4] They attend to the peculiar qualities of this familiarly mutable substance in order to raise questions about passion, sexuality and the human body.

There is a playful correlation between human bodies and coastal sands in the ways characters engage with the seaside environment in *Sons and Lovers* and *Maurice*. The protagonists' observations of and interactions with sandscapes appear to affect the way they understand themselves and the relationships with other characters both intuitively and in moments of reflection. In her essay "Nature's Queer Performativity", Karen Barad questions binaries such as those established between nature and culture, and the human and the non-human, by suggesting that human identity, including queerness, resides in the nature of what she calls *spacetimemattering*. Her neologism here expresses a concept which suggests that space, time and matter are not separate entities but are entangled.[5] In other words, we should not think about the world around us as divided into fixed and recognised categories, but as intertwined, just like the seaside: an in-between space that is ceaselessly moving, shifting and becoming. Barad argues that we need to

find ways to think about the nature of causality, origin, relationality, and change without taking these distinctions to be foundational or holding them in place. What is needed then is a way of thinking about the nature of differentiating that is not derivative of some fixed notion of identity or even a fixed spacing.[6]

Phenomena, including human identity, is the result of intricate entanglements of space, time and matter. Identity is not individual, but multiple, dynamic and ever-changing. Through such a rejection of the nature/culture binary, recent scholars have attempted to move away from thinking about queer theory only in relation to culture in order to find ways in which queerness can be associated with all kinds of natural processes.[7]

When writing about sand, foam and waves in the seaside, Lawrence and Forster wash away those binaries between humans and non-humans. Through attending to the temporariness of sand—a matter that is both natural and man-made, ephemeral and in the making—they show us ways through which we can understand human identity, body and emotions in relation to the transient and short-lived phenomena that happen on the beach.

* * *

The wide sweep that sandscapes present can give the impression to any onlooker that sand is all the same, in its creation of a false display or façade of homogeneity, but each grain holds its own unique features. It is possible, by looking closely through a microscope, to tell the origins of a grain of sand and, sometimes, the history of its journeys. Sand defies being categorised. Its unsettledness is also apparent in its association with time—sand grains gain their unique qualities over a long period of time, and sand is used to tell time in the case of the hour-glass. The verbal and embodied interactions that take place on the sands in *Maurice* and *Sons and Lovers* draw on analogies between the dynamic, temporal and indeterminate qualities of sand and the authors' understanding of the ambiguity, instability and unpredictability of passion and sexuality. Lawrence and Forster employ the meanings which the material qualities of sand generate to encourage multiple, unfettered interpretations of human identity and to destabilise fixed and normative concepts about sexuality.

Sons and Lovers is a *bildungsroman* of sorts and charts the story of Paul's sexual development. Emotionally influenced and attached to his mother, especially after the death of his older brother William, Paul develops his understanding of passion and sexuality during his sequential relationships with two women over the course of the novel. Lawrence depicts the seaside environment at different parts of the text in a way that reveals Paul's character. The openness of the sandscapes becomes analogous to Paul's free nature. We learn that as he walks on the sand towards Theddlethorpe in Lincolnshire, looking out to the sea, that Paul "loved to see it clanging at the land. He loved to feel himself between the noise of it and the silence of the sandy shore".[8] His appreciation of the seaside environment is physical and sensory, and while he recognises the clamour of the sea, he is positioned, and thereby associated with, the stillness of the sandscape. His attentiveness reveals his receptivity to the sound and touch of the material world around him.

Lawrence's interest in touch and the body has been noted by Abbie Garrington, who explains how he provides in his writing "a truly corporeal corpus, deeply invested in the experiences of the somatic system, and the philosophical and spiritual insight which consideration of the human body may bring".[9] Lawrence's emphasis on the role of the body in an individual's development demonstrates the ways in which bodily movements and inter-actions in the world have an impact on the mind, namely on perception and development. Paul's body, however, appears restrained during his walk with his girlfriend Miriam, a spiritual and religious young woman who holds dear her chastity and idealism. They both appear alienated from the environment around them, and the dark and secluded beach turns out to be more oppressive than freeing:

> The way home was through a gap in the sandhills, and then along a raised grass road between two dykes. The country was black and still. From behind the sandhills came the whisper of the sea. Paul and Miriam walked in silence. Suddenly he started. The whole of his blood seemed to burst into flame, and he could scarcely breathe. An enormous orange moon was staring at them from the rim of the sandhills. He stood still, looking at it. [...] He turned and looked at her. She stood beside him, for ever in shadow. Her face, covered with the darkness of her hat, was watching him unseen. But she was brooding. She was slightly afraid – deeply moved and religious. That was her best state. He was impotent against it. His blood was concentrated like a flame in his chest. But he could not get across to her. There were flashes in his blood. But somehow she ignored them. She was expecting some religious state in him. Still yearning, she was half aware of his passion, and gazed at him, troubled.[10]

The couple's emotional conflict on the sands is incompatible with their spontaneous surroundings. Paul's unplanned desire seems to be associated with and informed by the sands. Sandscapes, as places associated with illicit relationships and disapproved behaviours, seem to act upon Paul's other-wise repressed passion. Yet he feels compelled to ignore his bodily urges by distracting himself through looking at the distant and aloof moon instead, for "[t]he fact the he might want her as a man wants a woman had in him been suppressed into shame".[11]

Despite the private and sheltered atmosphere that the beach offers—there is the possibility here that he could give way to his feelings without public shame—Paul appears displaced physically and emotionally. By looking at and talking about the moon, and later shifting his vision towards the "one light in the darkness, the window of their lamp-lit cottage", he distances himself from Miriam, seeking shelter in his home and with his mother.[12] The beach here acts to expose Paul's feelings that, through his lack of physical interaction with the sandscape and with Miriam, become merely an oppressive and a limiting reminder of his body's unfitness to his society and he feels he must find (a quasi-) freedom somewhere else. Paul's body is exposed as unfitting on the sandy environment of the shore.

Miriam's body is also detached from her surroundings. She does not interact physically with the sand or with Paul. Instead, "she shrank in her convulsed, coiled torture from the thought of such a thing".[13] Her body responds to her mind and idealism to the extent that:

> this "purity" prevented even their first love-kiss. It was as if she could scarcely stand the shock of physical love, even a passionate kiss, and then he was too shrinking and sensitive to give it. As they walked along the dark fen-meadow he watched the moon and did not speak. She plodded beside him. He hated her, for she seemed in some way to make him despise himself. [...] Paul hated her because, somehow, she spoilt his ease and naturalness. And he writhed himself with a feeling of humiliation.[14]

Miriam's idealism does not only guide her actions but it also extends to Paul. By using quotation marks for the word "purity", the text questions this concept, which celebrates perfection and idealism and sets them in opposition to immorality. Purity is also the antonym for contamination, this sense of being in contact with something that is not just tainted but also unwanted. The juxtaposition of Miriam's purity and the sandy environment is ironic, as sand is a heterogeneous mixture of grains of different sizes, properties and sources. The moon, romanticised and far away, becomes an emblem signifying Paul's escape from their awkward silence and emotional reticence. Idealism and chastity are at conflict here with Paul's spontaneity, ingenuousness and openness. Miriam's claim to purity simply spoils the easy atmosphere of the seaside.

* * *

Passion is spontaneous. And, according to Clive in Forster's *Maurice*—as quoted at the opening of this essay—passion is sand. Sand grains assemble and move together, spontaneously, either as part of natural or man-made processes. Sand may come from rocks broken down slowly by other rocks to smaller pieces, while other grains of sand are ground by larger and smaller waves on the shore. Sand may also be the result of beach nourishment, a process in which sand is transported from other places to supply eroded beaches. In Jorge Luis Borges' poetry collection *In Praise of Darkness* (1969) he writes, "Nothing is built on stone; all is built on sand, but we must build as if the sand were stone".[15] Stone epitomises endurance but sand's transience more truthfully represents the natural process of things.

Sand, as Steven Connor elegantly puts it, "is the ultimate mockery of the permanence of stone, for it is no more than one of stone's own moods, the manner in which stone, atomised, consumes itself".[16] Sand is allied with the natural: impulsive, immediate and unrestrained. Sand's surfaces may mislead us to think that they are passive and inert, because they appear homogenous and uniform, but their agency materialises as a result of the forces of water and wind. Sand is as spontaneous as passion, a fact these novelists dramatise in their writings. In the beginning of *Maurice*, Forster depicts the different types of sand's natural movements, which challenge presumptions about the passivity of the seaside environment. He begins the novel with a provocative encounter that later determines Maurice's adolescent understanding of passion and sexuality, which reveals the ways in which the dynamicity of sandscapes can blur, confuse and question societal standards.

Maurice was published only a year after Forster's death in 1971. Its explicit depiction of homosexuality would have led to his prosecution had it been published in 1914 Britain.[17] The novel, a *bildungsroman* like *Sons and Lovers*, is concerned with the development of Maurice, especially the discovery, understanding, and acceptance of his sexuality. Just before progressing to his public school, fourteen-year-old Maurice has a private conversation with his teacher about the world of adults, the public school he is moving to next, and, eventually, women and marriage. Learning that he is the only man in his immediate family, Mr Ducie takes it as his manly duty to explain to the fatherless boy "the mystery of sex" before he hears about it from other boys. The teacher first speaks about males and females, illustrating his talk by scratching some diagrams on the sand[18]:

> Mr Ducie got up, and choosing a smooth piece of sand drew diagrams upon it with his walking stick. "This will make it easier," he said to the boy who watched dully: it bore no relation to his experiences. He was attentive, as was natural when he was the only one in the class, and he knew that the subject was serious and related to his own body. But he could not himself relate to it, it fell into pieces as soon as Mr Ducie put it together, like an impossible sum.[19]

Mr Ducie uses the convenient, flat surface of the sand for his lesson. He finds the subject matter, sex, difficult to express verbally, and so he chooses to draw it instead. Scratches on the sand, however, do not last. These short-lived and temporary marks will be changed and transformed before long. Unlike the durability and persistence of a slate or even paper, sand's granular qualities mean it is as crumbly and unreliable as brown sugar.

Although we never know precisely what the teacher draws, it is clear that Mr Ducie's illustrations attempt to explain the normative understanding of sexual relations between a man and a woman. By using a stick to illustrate his ideas instead of using his own fingers, Mr Ducie distances himself from the sand, which functions only as a blank and passive surface, useful for explaining this socially acknowledged understanding of sex. In this way, the stick becomes representative of hierarchical notions of normativity, demonstrating a superiority that attempts to impose only authorised scripts upon passive respondents. Maurice and his body in their turn may here act like "the smooth piece of sand", obedient and yielding to Mr Ducie's lesson. Yet, Maurice's lack of understanding makes it apparent that he will not be able to fully embody society's standards of masculinity and sexuality. These standards are outlined clearly to the reader in Mr Ducie's earlier conversation where he "spoke of the ideal man – chaste with asceticism. He sketched the glory of Woman. Engaged to be married himself, he grew more human, and his eyes coloured up behind the strong spectacles; his cheek flushed".[20] The chastity and self-discipline that Mr Ducie celebrates align with the purity that Miriam upholds during her relationship with Paul. These, however, are not concepts that are embraced by the sandscape.

In this beach scene, Maurice is still young and has not thought much about the topic of marriage or sex—or at least not how it is presented by his schoolmaster—and becomes only more puzzled when Mr Ducie's drawings prove to be temporary and unstable. The seaside turns into a place where his knowledge about sex remains unclear and his feelings become more, rather than less, confused. As they walk further,

> Mr Ducie stopped, and held his cheek as though every tooth ached. He turned and looked at the long expanse of sand behind. "I never scratched out those infernal diagrams," he said slowly. At the further end of the bay some people were following them, also by the edge of the sea. Their course would take them by the very spot where Mr Ducie had illustrated sex, and one of them was a lady. He ran back sweating with fear. "Sir, won't it be alright?" Maurice cried. "The tide'll have them covered by now." "Good Heavens…thank God…the tide's rising." And suddenly, for an instant of time, the boy despised him. "Liar," he thought. "Liar, coward, he's told me nothing."[21]

Beach sand is an unreliable surface not only because it is vulnerable to human manipulation but under constant interference from water and wind. Once observed by others, especially women, the diagrams turn from something which is supposedly normal and educative into something demonic ("infernal"). For a reason not made explicit in the text, Maurice begins to hate his teacher because he has been told "nothing". This sandscape scene of anger might be usefully compared to Lawrence's similar scene of sudden passion between Paul and Miriam. Paul's hatred for Miriam emerges from a sense of humiliation because "she spoilt his ease and naturalness". While the passage depicting Maurice's anger leaves the reader puzzled, the sand's association with spontaneity and freedom may explain the frustrations evident in both scenes. Maurice feels momentary anger when the sketches are deconstructed, following the inevitable shifting of the grains of sand on the beach. It is not only the drawings that disintegrate. Mr Ducie's idealism, self-discipline and abstinence from sensual pleasures become questionable when he loses his composure—he is no ascetic, passionless man after all. Static ideas about propriety and assertions of control are challenged by the disintegration of sandy images that affirm inert and static ways of thinking. In Forster's novel, the human body and sexual identity are associated with the unboundedness of the sand and the seaside environment, thus questioning idealistic social expectations. Sex is no longer a monolithic notion but is revealed to be as uncertain and indefinite as passion—or as sand. The mobile and unstable sandscapes depicted by Lawrence and Forster are not home to idealism or solid understandings of the world.

* * *

Sand may not possess sentient agency like living organisms such as crabs, anemone or lugworms, but the vivid description of the sands in *Sons and Lovers* and *Maurice* challenges humanist priorities and undermines rigid ways of conceptualising human sexuality, emotions and identity. Lawrence and Forster avoid presenting sand as a particular, predetermined entity by allowing sandscapes to actively emerge in the storyline and to shape the events, the identities and the development of the characters. Depicting decisive moments of Paul's and Maurice's sexual development on sands and associating them with the constant movement of seaside elements means that while their unboundedness and indecisiveness may be suppressed, limited and deceived by societal expectations, they are nonetheless defined by the dynamic material environment of the seaside.

It is worth noting here the second scene in *Sons and Lovers* where sands perform an active role in the relationship between Paul and his new lover, Clara, as well as the protagonist's understanding of passion and sexuality. Unlike his relationship with Miriam, Paul's intimacy with Clara is passionate and embodied, and situated on the beach where they both "went out together to bathe" and they "walked hand in hand".[22] Physicality is not only associated with bodies but with the seaside environment itself. Sand offers an environment of physical freedom which vitally informs their connection. We read that Paul "ran the cold sand through his fingers" and later, we learn of Clara "fingering the sand" as they discuss and contemplate the future of their relationship.[23] This proximity to sand is integral to their relationship as their bodies entangle with the vibrant materiality of the shore.

Barad's *spacetimemattering* is at work here. Though Paul and Miriam's bodies were subjected by obedience to traditional notions of chastity and idealism, the bodies of Paul and Clara respond to the elements of the seaside: "They shuddered with cold; then he raced her down the road to the green turf bridge" and "Clara stood shrinking slightly from the touch of the wind, twisting her hair".[24] Temperature and weather are presented as intimate companions of the lovers when they are on the beach. The repetition of the verb "shrinking" in the two scenes invites a comparison of the ways in which Miriam's and Clara's bodies respond to their surroundings. Clara is more responsive to the elements that the seaside environment offers, unlike Miriam whose body conforms to a sense of obligation towards certain religious and societal standards.

Paul's sensuous relationship with Clara corresponds to the materiality of the seaside environment and becomes more prominent in this scene. While the moon was distant during Paul's walk with Miriam, here, "the wan moon, half-way down the west, sank into insignificance. On the shadowy land things began to take life, plants with great leaves became distinct" and "The seagrass rose behind the white stripped woman".[25] If we take the moon to be indicative of social convention and a symbol of passivity, its withdrawal is noteworthy. The objects that are part of the landscape, "leaves" and "seagrass", become more noticeable and significant. His greater attentiveness to the seemingly irrelevant and trivial objects of the natural world reveals Paul's capacity for sensuous appreciation rather than, as critics have argued, any dismissal of human life. A case in point is W. H. Auden's observation that "Lawrence possessed a great capacity for affection and charity, but he could only direct it towards nonhuman life" and he "forgets about men and women with proper names and describes the anonymous life of stones, waters, forests, animals, flowers".[26] Lawrence's sympathy towards the non-human is indeed highlighted in this scene, but it is not just to displace the human, but to meditate on the intimacy between the human and non-human world as an indicator of the sensuality of human relationships.

The portrayal of Paul and Clara on the beach veers towards identifying them as part of the landscape. They are depicted almost as existing alongside the sand on which they lie, resembling a notion of human–non-human correspondence explored by Jane Bennet in her celebrated book *Vibrant Matter*. Here Bennett notes "the extent to which human being and thinghood overlap, the extent to which the us and the it slip-slide into each other". She reminds us that "we are also nonhuman and that things, too, are vital players in the world".[27] Human bodies and sands indeed overlap in Lawrence's writing. As the scene with Paul and Clara goes on, there is even a sense that the sandscape dominates and displaces the two characters.

The seaside's materiality becomes more vibrant in this part of the novel and has a direct impact on Paul and his emotions. His attention to the physical and the non-human even grows to the degree that the human recedes, losing agency and priority. Sand dominates the scene until Clara loses significance and identity to the extent that Paul cannot recognise her anymore. Like a grain of sand, she dissipates into her environment:

> He, on the sand hills, watched the great pale coast envelop her. She grew smaller, lost proportion, seemed only like a large white bird toiling forward. "Not much more than a big white pebble on the beach, not much more than a clot of foam being blown and rolled over in the sand," he said to himself. She seemed to move very slowly across the vast sounding shore. As he watched, he lost her. She was dazzled out of sight by the sunshine. Again he saw her, the merest white speck moving against the white, muttering sea-edge. "Look how little she is!" he said to himself. "She's lost like a grain of sand in the beach – just a concentrated speck blown along, a tiny white foam bubble, almost nothing among the morning. Why does she absorb me?"[28]

The sandscape here encompasses Clara until she becomes part of it, her body swallowed up by the mass of coastline. She becomes a temporary white spot, hardly identifiable as different from the sandscape around her. Connor writes about this quality of sand to overwhelm: "Sand has the capacity to engulf and inundate, blearing contours, eroding and erasing every edge and eminence".[29] Indeed, the magnitude of the coast alters the way Clara's form appears when seen from a distance, and the further she moves into the sandscape the more she becomes part of it. Clara becomes like sand and foam bubbles, temporary and fleeting. Sea foam is the result of transitory yet persistent tension, of agitation and turbulence in the different waves and layers of water that contain various types of organic matters. This temporary agitation is similar to passion, an impulse and an urge. Not enough, according to Clive in *Maurice*, to build a relationship on.

* * *

When exploring the permanent and temporary qualities of the seaside, Khalil Gibran writes in his collection of inspirational quotes, *Foam and Sand* (1926):

> I AM FOREVER walking upon these shores,
> Betwixt the sand and the foam,
> The high tide will erase my foot-prints,
> And the wind will blow away the foam.
> But the sea and the shore will remain
> Forever.

Although concluding with the permanence and infinity of the sea and the shore, the above lines emphasise the temporality of human interaction with sand and foam. Writing about Paul's desire towards Clara reminds the reader of more than just human humility. The image is a reassertion of the immediacy of human identity and emotions, as well as their constant transformation.

Paul is mystified by his attraction to Clara, for he asks "why does she absorb me?". It is as though he is the water, unavoidably pulled into the lattice-like structure of grainy sand. Later, his musing returns him to the image of foam:

> "What is she, after all?", he said to himself. "Here's the seacoast morning, big and permanent and beautiful; there she is, fretting, always unsatisfied, and temporary as a bubble of foam. What does she mean to me, after all? She represents something, like a bubble of foam represents the sea. But what is she? It's not her that I care for".[30]

Clara is depicted as a part of something bigger, a temporary foam in the midst of an eternal sea. Seeing her engulfed in the vastness of the waterscape affects Paul and how he perceives their relationship. Clara is displaced by her non-human surroundings, first as a pebble, then foam and later a grain of sand. Paul's perception of Clara, unwittingly perhaps, reveals her fleeting status in his heart, and the transitoriness of human emotions per se. When writing on sand in his novel, Lawrence shows how the material world has an agency or life of sorts and asserts that inert entities have their own ontologies that are separate from human consciousness. In *Sons and Lovers*, Lawrence's empathetic descriptions and treatment of sand and his meticulous readings of the characters' emotional responses to it enable us to pay attention to the way sand operates as a catalyst in Paul's sexual identity.

* * *

The ways in which the characters interact with the coastal environments in *Sons and Lovers* and *Maurice* reveal a remarkable engagement with and an empathetic attitude towards not those seaside characteristics normally celebrated, such as air, the sea, the atmosphere or leisure, but rather sand and its transformative qualities. The two Edwardian writers contemplate the ways in which sandscapes decentre the human by bringing both the human and non-human closer together, undermining the centrality of human rational agency.

The seaside offers the writers a place to celebrate the dynamicity and unsettledness of what it means to be human. Their writing shows, what Lawrence tried to explain in numerous essays, how "Each thing, living or unliving, streams in its odd, intertwining flux, and nothing [...] is fixed or abiding. All moves".[31] Nature is always in process, constantly making and unmaking, growing and decaying, attracting and repelling. What Barad identifies as natural, as I expressed earlier in the chapter, is not what the term has been used for traditionally. The natural for Barad is not that which is predetermined, repetitive or follows a clear course of action—that which follows a cause and effect logic—but instead that which is contingent and happening; for "all bodies, not merely human bodies, come to matter through the world's performativity – its iterative intra-activity".[32] It is important to attend to the implications of twenty-first-century studies, such as those by Barad, which understand human identity, body and sexuality in relation to natural processes and as bounded to *spacetimemattering*, because they help us to understand the ways in which Lawrence's and Forster's creative fiction engages with the performances and processes of the natural world, the human body and sexuality alike.

Sand is integral to the activities—walks and conversations—that take place on the seaside. It exposes the proximity of the material environment and characters' bodies, and the mutual development of identities and sexualities. To Lawrence and Forster, passion is indeed sand. Its immediacy, spontaneity and unpredictability move it outside the stiff and fixed categories of social conventions. Like sand, passion refuses to be contained, defined and categorised. Sand's granularity challenges perceptions of the seaside or of human identity as solidified or contained. Sand is not just an inert backdrop but is vibrant, and in a limited sense, alive.

Notes

1. E. M. Forster, *Maurice* (London: Penguin Group, 2005), 113.
2. See Howard J. Booth, "D. H. Lawrence and Male Homosexual Desire", *The Review of English Studies*, 53, no. 209 (2002): 86–107 and Stephen Da Silva, "Transvaluing Immaturity: Reverse Discourses of Male Homosexuality in E. M. Forster's Posthumously Published Fiction", *Criticism*, 40, no. 2 (1998): 237–272.
3. Daniel Burdsey, *Race, Place and the Seaside: Postcards from the Edge* (London: Palgrave Macmillan, 2016), 43–44.
4. Fluidity is a term that is primarily used in discussions of bisexuality. Clare Hemmings writes how "Rather than identifying as homo or hetero, bisexuals find themselves in an essential state of FLUIDITY and NON -IDENTITY", in *Bisexual Spaces: A Geography of Sexuality and Gender* (New York: Routledge, 2020), 173. The same term, however, is often used in looser terms to indicate postmodern understandings of non-normative sexuality. Esther Rapoport explains how "queer theory emphasizes the fluidity of human sexuality and in certain ways establishes fluid eroticism as an idea", in "Bisexuality in Psychoanalytic Theory: Interpreting the Resistance", in *Bisexuality and Queer Theory: Intersections, Connections and Challenges*, ed. Jonathan Alexander and Serena Anderlini-D'Onofrio (Abingdon: Routledge, 2012), 99.
5. See Karen Barad, *Meeting the Universe Halfway: Quantum Physics and the Entanglement of Matter and Meaning* (Durham: Duke University Press, 2007).
6. Karen Barad, "Nature's Queer Performativity", *Qui Parle*, 19, no. 2 (2011): 121–158 (124).
7. See Timothy Morton, "Guest Column: Queer Ecology", *PMLA* 125, no. 2 (2010): 273–282.
8. D. H. Lawrence, *Sons and Lovers* (Cambridge: Cambridge University Press, 2013), 220.
9. Abbie Garrington, *Haptic Modernism: Touch and the Tactile in Modernist Writing* (Edinburgh: Edinburgh University Press, 2015), 155.
10. Lawrence, *Sons and Lovers*, 220–221.
11. Lawrence, *Sons and Lovers*, 221.
12. Lawrence, *Sons and Lovers*, 221.
13. Lawrence, *Sons and Lovers*, 221.
14. Lawrence, *Sons and Lovers*, 221.
15. Quoted in Vince Beiser, *The World in a Grain: The Story of Sand and How It Transformed Civilization* (New York: Riverhead Books, 2018), 27.
16. Steven Connor, "The Dust That Measures All Our Time" (May 2010), http://stevenconnor.com/sand/.
17. Prosecution would have been under what is called the "blackmailer's charter", as David Leavitt explains in one of his notes in the introduction of *Maurice*: "Labouchere Amendment of 1885; according to its provisions, 'acts of gross

indecency' between adult men, in public or private, were punishable by up to two years in prison", xxxii.

18. Forster, *Maurice*, 8.
19. Forster, *Maurice*, 9.
20. Forster, *Maurice*, 10.
21. Forster, *Maurice*, 10.
22. Lawrence, *Sons and Lovers*, 438, 439.
23. Lawrence, *Sons and Lovers*, 442, 443.
24. Lawrence, *Sons and Lovers*, 440.
25. Lawrence, *Sons and Lovers*, 439–440.
26. W. H. Auden, "D. H. Lawrence", in *The Dyer's Hand and Other Essays* (London: Faber and Faber, 1961), 277–295 (289).
27. Jane Bennett, *Vibrant Matter: A Political Ecology of Things* (Durham: Duke University Press, 2010), 4.
28. Lawrence, *Sons and Lovers*, 440–441.
29. Steven Connor, "The Dust That Measures All Our Time", http://stevenconnor.com/sand.html.
30. Lawrence, *Sons and Lovers*, 441.
31. D. H. Lawrence, "Art and Morality" [1925] in *Phoenix: The Posthumous Papers of D. H. Lawrence* ed. Edward D. McDonalnd (London: Willian Heinemann Ltd., 1936), 525.
32. Barad explains: "The notion of intra-action (in contrast to the usual 'inter-action', which presumes the prior existence of independent entities/relata) marks an important shift, re-opening and reconfiguring foundational notions of classical ontology such as causality, agency, space, time, matter, discourse, responsibility, and accountability", in "Nature's Queer Performativity", 125.

7

On the Sound-Sea: Fifteen Ways of Thinking About Sand and Sound

Brian Baker

Sandscape

Where I grew up, on the north bank of the Thames estuary in Essex, "sand-scapes" and "soundscapes" are homophones. The long, lazy vowel in the Essex pronunciation of "sound" is like the slow slipping of the foot into silt and sand, or the filling of the footmark with salt water. The sound of the sea in Southend or Leigh-on-Sea is a gentle lapping, the wash of brownish water on tidal mud, rather than the exhilarating crash of surf on stones. At the river's mouth, the air is briny, with notes of fish and effluent, the particular stink of marsh and mudflat. In Southend, the soundscape is dominated by the cars that pass along the seafront, the gulls, people promenading or walking the dog, and towards the pier, the arcades and Adventure Island and Rossi's ice-cream parlours (Fig. 7.1).

Saturday

My grandfather Jim worked as a driver during and after the Second World War, taking essential supplies to and from the docks. He had the Londoner's love of the seaside and eventually moved with his sons (one of them my Dad) and his second wife to Essex in the late 1950s. Though always oriented back

B. Baker (✉)
Department of English Literature and Creative Writing, Lancaster University, Lancaster, UK

© The Author(s) 2020
J. Carruthers and N. Dakkak (eds.), *Sandscapes,*
https://doi.org/10.1007/978-3-030-44780-9_7

Fig. 7.1 *Southend.* Brian Baker, monotype, 2019

towards North London, an angel of history passing down the A13 backwards, he drove down to the Southend seafront as often as he could. Regularly, he and my Nan would pick me and my sister up in their Ford Cortina to drive us down to have a stroll along the front and an ice cream at Rossi's. The sound of Jim telling stories or Nan's good-natured bickering would track a few hours on a Saturday afternoon, before we'd drive home for tea.

Ten to five, the teleprinter, the football results, the Pools. The Pools, a weekly gamble based upon predicting football scores, was run largely by Zetters and Littlewoods, the latter based in another 'Pool—the city of Liverpool. My Nan and Jim played every week but barely won back their stake. A pool of hope, consisting of a million players, trying their predictive skill against the bounce of the ball, the weather, bad refereeing decisions, good or poor play, and the changing form of the teams. Standing in concrete bowls, the sound of the crowd in a football stadium is a reverberation of fifty thousand voices joined together, pooling the sound, channelling it and circulating it. The deep boom of voices is like the sound of the sea against the rocks, the crash and echo of a million droplets of water striking stone.

Saturday afternoon was the time for leisure. With another day of rest to follow, there was no hurry. A stroll along the seafront, some window-shopping, a cup of tea in a café and a lemonade for the kids. Something special for tea, maybe. The sound of leisure was slow time, voices in conversation, catching a little of radio commentary on the cricket, a pause to eat a foil-wrapped sandwich. There's no hurry.

Silt

Robert Macfarlane's "Silt" in *The Old Ways* (2012), about the Broomway that runs out onto the sand further east towards the estuary's mouth, at Foulness, focuses on the emptiness, the eerie shimmer of water on mud and sand. He only interacts with the inhabitant of the island who invites him on to Foulness by letter, not in person (access is restricted because of the military firing ranges, the booms which you can still hear from my parents' house when the wind is from the east). Out there, beyond Shoeburyness, where the houses dwindle into the marsh, the soundscape empties of human cacophony—unless you bring it with you, of course.

There is very little sound in "Silt". As Macfarlane and his walking companion leave Foulness island, they hear three oystercatchers as they dart overhead; distant gunfire and a cuckoo; and out on the sands, they "could hear the man whistling to his dog, now far away on the sea wall. Otherwise, there was nothing except bronze sand and mercury water".[1] As the walkers enter into a depthless, scale-distorting flatness, sound seems to disappear. The writing is intensely visual, visuality leading into interiority, their steps leading both out and in. There is a curious element to the whole walk, however. Warnings and prohibitions are everywhere, from embarking on the walk itself to taking photographs of MoD land. The intertidal space beckons, but is dangerous, as walkers may lose the path, and as Macfarlane notes, the tide can come in quicker than a human being can run. But there is something else, a desire to go out further, to lose the path. As they set out on the walk, Macfarlane notes that "I subdued the alarm my brain was raising at the idea of walking out to sea fully clothed, as only suicides do"[2]; but on the way back he feels "a powerful desire to walk straight out to sea and explore the greater freedoms of this empty tidal world".[3] On the return journey, they do just that, walking off the path and exploring the flats: "It felt at that moment unarguable that a horizon line might exert as potent a pull upon the mind as a mountain's summit".[4] Although the knowledge of the tide's return, half-suppressed in the shifting sandscape, eventually seeps back into their minds, propelling them back towards shore, there is a longing here, a longing to remain in the empty world. If we creatively conflate the two sentences I quoted above, Macfarlane experiences "a powerful desire to walk straight out to sea [...] fully clothed, as only suicides do".

Speedos

The music of the Essex seafront is fast electric rhythm and blues, as characterised in the film *Oil City Confidential*, about the Canvey Island band Dr Feelgood.[5] Their gig at the Kursaal, a domed venue on the front, is the stuff of legend. Today, Eight Rounds Rapid continue this tradition, with the son of Wilko Johnson (the original Feelgood's guitarist) flashing his choppy guitar riffs and the lead singer purveying a none-more-Essex vocal style. Their single "Channel Swimmer" lists methods of failed and successful suicides: a car pile-up, the dry swallow of pills, the "swimmer" themselves.[6]

The video was shot on an empty Southend beach, the band clad in Speedos. It begins without music, as three swimmers and their trainer walk in silence down a concrete runway to the beach, where a sharp cut away from their faces reveals the tide is a long, long way out. The band then strike up with a fast, jagged guitar riff, the band now playing their instruments on the beach, all but the guitarist/trainer in Speedos, swimhat and goggles. The singer leans right into the camera, his long face somehow as rubbery as his hat in the black-and-white footage. Later in the video, the band wander along the Southend seafront, play on the penny arcades, and have a beer outside a pub. At the end of the song, the three in swimsuits run off and dive into the water, the tide now back in and, as the trainer watches, seem to disappear like the Channel Swimmers they sing about.

Shellfish

Jim had a Londoner's love of shellfish: cockles, whelks and mussels bought from an Old Leigh stall, liberally soused in malt vinegar. A jellied eel was not unknown. Not oysters, though—these had long passed upmarket. Up the Essex coast, at the marshes at Tollesbury, my forebears on my mother's side worked as oyster fishermen off the mouth of the rivers Crouch and Blackwater. The Carters fished for oysters for generations as cousins worked on the land. I still have the salt water in my veins, as well as the labourer's spade-like hands.

In "Silt", Macfarlane notes that "the Broomway is the less notorious of Britain's two great offshore footpaths, the other being the route that crosses the sands of Morecambe Bay".[7] That's not the only thing that connects the two spaces, Morecambe and the South East Essex coastline. The respective football teams are known as the Shrimps and the Shrimpers, for instance; for the Bay and the Midland hotel, for Southend read the estuary and the Pier. In Justin Hopper's excellent site-specific work *Public Record*, recorded with a soundscape by Scanner (Robin Rimbaud), he describes the loss of life to fishermen on their trawlers, accidents when fishing smacks were run down by large merchantmen heading for the Pool of London.[8] This part of the Thames is a busy sea lane now, huge container ships passing by on their way to the giant cranes at the Tilbury port.

Macfarlane also notes the death of the Chinese cockle-pickers on Morecambe Bay in 2004, a stark reminder that sand is also a place of work as well as of leisure, a place of danger as well as pleasure. In North London, when Jim was young, Sunday teatime would be punctuated by the call of the fishmonger, selling half-pints of cockles, whelks and winkles in their shells for high tea. From the Thames estuary, where the boats would bring in their "catch" of shellfish, to the tables of Camden Town, a taste and smell and sound of the sea.

Silence

I stood on the beach at Newborough, on Anglesey, on a grey afternoon. We had been there as a family many times. The beach has its own micro-climate, caused by the passage of the water through the Menai Straits and the presence of Snowdonia, looming behind Caernarvon, where clouds bubble up as the heat rises from the mountains. The beach is wide and curved and, when the tide goes out, you can walk across to a small island where there stood a lighthouse and is now a nature reserve. On the other side of this tidal isthmus is an even wider curve of sand that tends to be much quieter, with fewer dog walkers and families. The sand is hard and compacted when the tide goes out, and the undulations of the beach's topography form pools that slowly drain, leaving islands and then expanses of rippled sand. I have been there in all seasons, with high tide and low, with the wind whipping sand off the dunes and the kite-surfers zooming along the waves, with a soft rain falling, and with the sun glittering from the surface of the sea. I have recorded my young daughter shouting "splish, splash" as she jumped over the small waves as they came in, holding on to my hand. I have walked here with my wife, both bundled up against the cold, my ears envying her woollen hat (Fig. 7.2).

Fig. 7.2 *Newborough.* Brian Baker, pastel on paper, 2019

Sea and Sand and Stones

Analemmatic clocks, made by my wife Deniz from rounded slate shards, tell the time of the beach. Stones form lines, traps for toes. The sound of a quickly in-drawn breath. A stone beach has a sound of its own: a boom as the wave descends, then a "shh" as the sea recedes, accompanied by a million knocks of rock on rock, the workings of a geological clock as the stones roll, pulled back down the beach. Sand is weathered stone, stones at the granular level, their hiss at frequencies the human ear cannot catch. The stone beach, every few seconds, presents to the ear the workings of a sidereal mechanism that tolls the cosmological hour. It is the earth's clock, the moon's clock, the star's clock.

In the story told by The Who's album *Quadrophenia*, Jimmy, a young Mod, suffering from a personality disorder that Pete Townshend dubs "quadrophenia", has a kind of breakdown, and travels from his South London home to Brighton, where he has an epiphany at the seaside.[9] The first song on the double-album, "This Is the Sea", is a soundscape: amidst the sonic wash of high waves crashing on to Brighton's stony beach, wisps of songs pass like ghosts, or tolling on the offshore breeze. Each of the musical motifs relates to one of the band members, four become one in the sea. The album ends ambiguously: the listener doesn't know whether Jimmy ends his life in the sea (one of the songs is called "Drowned") or whether he simply throws off the burdens of being a Mod. The double-album comprises four sides of music and a booklet of a textual narrative and a series of photographs illustrating Jimmy's journey. Jimmy feels the tension between wanting to be the "real me", an individual, with the pleasures of being a Face in the crowd as a kind of crisis or rupture.

Jimmy's journey to the sea can be thought of as an enactment of a desire to "drown" and dissolve the unsustainable fragmentation of quadrophenia in a "oneness" that is without boundaries altogether. Total dissolution, death by drowning, is implied in the shot where Jimmy is fully submerged under the water. This is not the last shot, however. Jimmy makes it to "The Rock", and the final shots in the booklet show him walking alone on the shore, half-in and half-out of the water. In these images, *Quadrophenia*, the album, rejects suicide as a means by which to transcend the disabling tensions produced by masculine subjectivity and the need to rupture it and to "explode" out of it. Instead, Jimmy maroons himself on another beach, walking the tideline, *between* the sea and the sand rather than *by* it. The imagery of the rock, phallically protruding from the sea but deeply invaginated, echoes this concept of the beach not only as the place where he feels "real", but also one where the constructions of gender are themselves in flux.

Sound/Less

The opening of Iris Murdoch's *The Sea The Sea* (1978) begins with the description of the coastline by a first-person narrator, whose eye is wonderfully painterly:

> The sea which lies before me as I write glows rather than sparkles in the bland May sunshine. With the tide turning, it leans quietly against the land, almost unflecked by ripples or by foam. Near to the horizon it is a lustrous purple, spotted with regular lines of emerald green. At the horizon it is indigo. Near to the shore, where my view is framed by rising heaps of lumpy yellow rock, there is a band of lighter green, icy and pure, less radiant, opaque however, not transparent.[10]

This is the beginning of a "memoir" which, we are told, has been inserted into the beginning of a "diary" or "chronicle" written by the narrator: a more formal piece of writing, a set piece, transplanted into the start of something more informal. What is noticeable about the opening is its visual power, its intention to make the reader *see*: what we do not find here is *sound*. This is narrative description as a painting, a visual spectacle rendered into language. It depicts the scene marvellously but is distanced from it because there is no soundscape, no immersion in the world. We could be in a gallery describing a painting.

By comparison, Cynan Jones's *Cove* (2016), a novella that narrates the experience of a fisherman struck by lightning while at sea, begins: "You hear, on the slight breeze, the *tunt tunt, tunt tunt* before you see the boat. You feel illicit".[11] This first section is narrated by the fisherman's pregnant wife, waiting on the beach for her husband's return. On the beach, she sees the doll, washed up, of a missing child; later in the novella, but earlier in time, the fisherman also encounters the doll, a rather over-determined symbol of loss and trauma. When the lightning strikes, sound is as important as vision:

> The wind picks up, cold air moving in front of the storm.
> And then there is a basal roll. The sound of a great weight landing. A slow tearing in the sky.
> One repeated word now. No, no, no.
> When it hits him there is a bright white light.[12]

In this passage, the flash of lightning, the intense moment of vision that ends vision (as it shocks him into unconsciousness) closes the world of sound, completes it. The wind, the thunder, the words: and then the strike.

Surf

In *The Soundscape*, Murray Schafer writes: "What was the first sound heard? It was the caress of the waters".[13] But the sound of the sea can be more violent, more overwhelming than a caress. The overwhelming sound of the sea is surf: high, white and green. Breakers grasp and sting the slate of Llangrannog on the Ceredigion coast of Wales, cliff faces etched black by salt and spray. Sand flies cling to legs, searching for wrack. On another day, jellyfish lay in terminal pools, transparent innards on show, an alien submarine squadron crashed to earth. Dogs scatter the sand as it spurts behind them as they sprint; or, they stand in the surf, backside to the sea, waiting (Fig. 7.3).

The hiss, the shell-roar, the deep cough of the waves expiring like radio static on the sand: an ocean of sound, a sound-sea. The notched slate tooth at Llangrannog, Carreg Bica, looms in my peripheral vision like someone sitting up on the rocks behind me, knees up, face jutting towards the horizon. As Joe Banks suggests in *Rorschach Audio*, my mind makes pattern from abstraction, meaning from meaninglessness. The metaphors I use for the sea also suggest a

Fig. 7.3 *Llangrannog*. Brian Baker, monotype and watercolour on paper, 2019

presence or voice: a hiss, a whisper, a roar. The sea possesses subjectivity; not Neptune, but a negotiation between the sea itself and the one who stands and hears. The tideline, moving up and down the strand, is a marker of this negotiation. The beach is an intertidal space, there and not there; the sound of the sea recedes, dims perhaps, twice during the day.

At the end of Jack Kerouac's *Big Sur*, there appears a poem in free verse called "SEA: Sounds of the Pacific Ocean at Big Sur". In it, the onomatopoeic qualities of language are used to bring forth the experience of being next to the sea, standing on the shore while the sound of the surf envelops the reader. The method is explicitly Joycean—later, we find the lines "Green winds on tamarack vines– / Joyce-James-Shhish– / Sea – Sssssss – see"—in a *Finnegans Wake*-emulating fullness of language, the rolling of the waves around the tongue.[14] There is something Babel-like in the acoustic qualities of the verse, language shifting from onomatopoeia ("roll, roll") to a neologism ("pali andarva") to a purely phonic rendition of sound ("Shish") to French ("parler"). It is as if the sea contains all languages, is the mother of languages ("speak you parler / in this my mother's / parlor"), and speaks in a multitude of tongues.[15]

The single individual, Kerouac's male seeker, encounters a feminine multitude, and wants to be part of the sonic conversation by emulating and transcribing it, yet reinforces separation from it. The persona turns his back on the sea, frightened by the waves and surf and sound, refusing communion with it. Instead, the male seeker turns within himself: "Not tempest as still & awful / as the tempest within".[16] The tempest within, in Kerouac's work, is of course represented by language, by the tumbling, ongoing streams of internal speech. The beginning of *Big Sur*'s first chapter indicates that there is no tension between the language of Kerouac's narrative prose and the method of "SEA":

> I wake up drunk, sick, disgusted, frightened, in fact terrified by that sad song across the roofs mingling with the lachrymose cries of a Salvation Army meeting on the corner below [...] and worse than that the sound of old drunks throwing up in rooms next to mine, the creak of hall steps, the moans every-where – including the moan that awakened me, my own moan in the lumpy bed, a moan caused by a big roaring Whoo Whoo in my head that had shot me out of my pillow like a ghost.[17]

Where the end of *Big Sur* is a kind of encounter with sound as other, even if one that is ultimately refused, here sound is infernal, torturing—and corresponds to the wreckage and wailing within. The trajectory of the novel is from one soundscape to another, as much as it is from the city to the sea: but peace, tranquillity, silence, is impossible to find or represent until the novel ends and we are left with the blank, white page.

Silence

I stood, on a grey March day, on the beach at Newborough. We had been there as a family many times. In the past few months, I had suffered what used to be called "a breakdown", but with help of family and doctors I had carried on functioning. I was suffering from—I do suffer from—depression, long undiagnosed. Both deep-rooted and more recent events had finally caused my coping mechanisms—to withdraw, to close down, to empty myself out, to become nothing—to finally malfunction. I had imagined driving to this beach, early in the morning, without telling anyone. I had imagined not coming back.

Starlings

At Aberystwyth, the starlings swirl in a large, ever-shifting flock above the pier, a few metres from the beach. This graceful, uncanny group motion is called a murmuration. The pulsing, fluid pattern suggests intelligence, a mind that co-ordinates the motion. The organic forms seem to coalesce, transform,

Fig. 7.4 *Murmuration*. Brian Baker, monotype and ink on paper, 2019

disperse and re-form in constant flux. Yet there is no call, no chatter as of swallows in flight. The starlings in their murmuration gather and fly to the hush of the sea and the call of the gulls. A helix, a tornado, a Miro, a Barbara Hepworth, dancing in the sky above the pier (Fig. 7.4).

Seabirds

I saw a photograph someone had taken while standing on a promenade. It was a close-up, the photographer's left hand clutching a double cheeseburger, which was in the process of falling apart. Above the burger bun stared the yellow eye of a gull, its beak speared through the bread in a contest for possession. The beak was open, and the look in the bird's eye was piratical. The photo appeared as a parody of those lovingly framed bistro-burgers impaled by a cocktail stick, a spiky spindle. This burger, however, would soon be airborne.

A gull is a seabird, but a gull is also a credulous fool. As a verb, to gull is to deceive, make a fool of.

The call of gulls, between a raven's caw and hyena's laugh, is a mockery. They wheel and eye the walking, wing-tips feeling for flaws in the wind. They sail, lift, screech. They insist the space is their own.

"The Seabirds" is the first song on The Triffids 1985 album *Born Sandy Devotional*.[18] The front cover of the album is an aerial shot of the coast of Western Australia, where the band were formed: in particular, Mandurah township, with aquamarine seas, sand the colour of ripening barley, and an inlet taking a stream to the sea. "The Seabirds", however, like several other tracks on the album ("Tarrilup Bridge", "Lonely Stretch") is about loss and death. It is a kind of grand, melodramatic ballad, given Country inflections by Graham Lee's swooping pedal-steel guitar work. The lead singer, David McComb, sings of a man whose relationship has failed and who finds himself on a beach, listening to the screams of the gulls as they pick at the eyes and bodies of the fish in the bay, turning the water red. The "devotional" in the album title gives a sense of the apocalyptic or even scriptural quality of the scene we are given; this is not quite the Gothic territory of Nick Cave and the Bad Seeds, but the two bands did have the same bassist, Martyn P. Casey.

The protagonist of the song swims out to the reef, the coral cutting his skin, another image of blood in the water. He offers himself to the birds, avowing that he is no longer afraid to die, but the gulls will not touch him, even when his dead body is washed up on the beach. Underpinning the song is a sense of isolation, that there was no one there to tell him that there was another path, and the song ends with an accusation, to the listener perhaps: Where were you? In "The Seabirds", the sand is a terminal space, haunted by the cries of gulls. But there is a sense here that this death was not inevitable, even if the gulls are indifferent.

Suicides

In the twelfth, thirteenth and fourteenth cantos of Dante's *Divine Comedy*, Dante Alighieri and Virgil approach the Seventh Circle, which holds the Violent. In the outer part of this circle are the violent against others, who are tormented in like manner. The middle part holds the violent against themselves, "and therefore in the second round must repent in vain whoever robs himself of your world or gambles away or dissipates his wealth, lamenting where he should rejoice".[19] In the inner part of the Seventh Circle are the violent against God, nature and against art. Between the second and third rounds, however, lies a boundary: "the ground was a dry, deep sand" populated by "herds of naked souls who were all lamenting miserably".[20]

In the middle, or Second round Dante encounters the Suicides. In a typi-cally striking image, the souls of suicides are flung by Minos to the seventh depth, where they land as seeds, randomly, and sprout "like a grain of spelt and rise to a sapling and a savage tree; then the Harpies, feeding on its leaves, cause pain and for the pain an outlet".[21] The Harpies themselves cry horribly, and the trees wail their lamentations. These souls or shades of the suicides take the form of gnarled, twisted, thorny trees which, like the one Dante found himself in before encountering the door to the Underworld, creates a trackless wood. The Promethean punishment, that the leaves grow to be pecked and eaten by the Harpies, indicates the extent of this trespass against God's will. At judgement day, one of the trees explains,

> Like the rest we shall go for the cast-off flesh we have left, but not so that any of us will be clothed in it again, for it is not just that one should have that of which he robs himself. We shall drag them here and through the dismal wood our bodies will be hung, each on the bush of its injurious shade.[22]

This odd echo of crucifixion, taking place for all eternity, is a harrowing prospect that leaves Dante speechless with pity. The orderless, trackless wood, filled with the lamentation of those whose suicide is a rejection of God's grace, borders the sand and is, in a sense, as lifeless and as tormenting as that scorching sand. Here, nothing grows, but as pain, a typically symbolic instantiation of what is assumed to be a perversion of the principles of life.

In *Notes on Suicide*, Simon Critchley traces Sigmund Freud's writing about depression in the 1917 essay "Mourning and Melancholia". "What happens in depression", Critchley writes, "is that the self turns against itself, the subject makes itself into an object, and complains bitterly".[23] Freud's point, he suggests, is that

> in order to kill ourselves we have to turn ourselves into objects. More precisely, we have to turn ourselves into objects that we hate. Thus, suicide is strictly speaking impossible. I cannot kill *myself*. What I kill is the hated object that I have become.[24]

This is Freud's point, Critchley argues: "Suicide is the determination to rid ourselves of what enslaves us: the mind, the head, the brain, that vague area of febrile activity somewhere behind the eyes".[25] This is why David Foster Wallace, in *This Is Water*, noted that people shoot themselves in the head, not the heart, Critchley explains.[26]

Suicide, then, is a form of rebellion, as Dante imagined. Where the medieval Florentine presented this in terms of God's will, Critchley (following

Freud) sees it as a rebellion against the rational subject, the thinking consciousness. It is an inversion of the *cogito*, or the *cogito* re-conceptualised through the death drive, where death is a return to quiescence, the end of suffering: not "I think therefore I am", but "I/you think therefore I/you must die".

Sirens

One of my favourite bands is the Cocteau Twins. Their music, with the chiming, reverberating guitar work of Robin Guthrie, and the soaring soprano vocals of Elizabeth Fraser, often forms soundscapes, particularly when drums and percussion are absent (as on 1987's *Victorialand*).[27] Their album with Harold Budd, the pianist who often plays "treated" piano, *The Moon and the Melodies* (1986) begins with "Sea, Swallow Me", a rare direct reference to the sea in this most oceanic of bands.[28] Perhaps Fraser's most well-known performance, however, was not with the Cocteau Twins, but with This Mortal Coil, a collective formed by the founder of the Cocteau Twins' record label, 4AD, Ivo Watts-Russell. She sings on a cover version of Tim Buckley's "Song to the Siren", from the point of view of a protagonist who has been "shipwrecked" in the ocean, and has been pulled to safety by the Siren, who sings and promises to enfold the sailor.[29]

The lyrics are clearly a metaphor for love, loss and desire as much as they tell a story, although there is an intense visuality which finds a strange counterpoint in Fraser's vocal delivery which, like on Cocteau Twins tracks, almost masks language and performs *song as sound*. Here, though, there is much more purchase on lyrical clarity than on "Sea, Swallow Me", as the singer stands among the breakers and wonders whether she/he should lie with "death, my bride". As sung by Tim Buckley, this is a song sung *to* the siren, as the title suggests; but sung by Fraser, this seems like the song *of* the siren, the gender reversal rendering the song opaque and ambiguous, if musically lovely. Rather than Buckley telling us, at second hand, of the Siren's song, Fraser enacts it, and offers the consoling, enticing sound of the Siren herself.

Silence

I stood on the beach at Newborough. My wife and two daughters were there with me. I stared at the sea, and they did not know what I was doing. My younger daughter spoke to me but I did not hear and, I was told later, I did not respond. Something within me, something that wanted to hold and protect me above all else, called to me. I wanted, very much, to walk directly into the sea, fully clothed. But I stayed, standing on the beach.

Sunday

Morrissey's second solo single, released in May 1988, was "Everyday Is Like Sunday".[30] Composed by Stephen Street in a jangling guitar-pop style that strongly recalled the sound of The Smiths, Morrissey's vocals are in the second person, singing to a "you" who lives in a declining seaside town, where nothing ever happens, the shops are all closed, and all you can do is wait to leave. There is an apocalyptic edge to the chorus, where it is called down almost as a relief from the boredom, to finish the job that the Luftwaffe or post-war "redevelopment" left incomplete. As usual, Morrissey's arch delivery provides a comic edge to the gloom, almost satirising the teenage angst that seemed to make up much of his constituency in the 1980s (including me). In the video, however, Morrissey hardly appears, except for in a waggish final shot, where he appears to be standing on the beach. For the foregoing three minutes, the video follows a day in the life of a young woman in just such a seaside town. It could be Southport, it could be Morecambe, it could be Skegness; but the video was shot, in fact, on the seafront of Southend-on-Sea. One short scene also features the young woman in a record shop: Golden Disc, which I recognised as I spent many teenage hours browsing there. (The other main record shop was Parrot Records in the "precinct" at the other end of the High Street, a windowless cave of a shop that contained many independent-label treasures.) At a time when opening hours were still heavily restricted in the UK, Sunday was not a time for browsing records, but for listening to them in your room.

Soundscape

In *The Soundscape*, Murray Schafer writes about *"keynote sounds, signals* and *soundmarks"*. The keynote sounds are "those created by [...] geography and climate"; "Signals are foreground sounds and are listened to consciously"; and soundmarks refer to "a community sound which is unique or possesses qualities which make it specially regarded or noticed by people in that community".[31] The soundscape, constituted by all the keynotes, signals and soundmarks themselves constitute place but also, in various ways, constitute subjectivity. In soundtracks, we Walk(man) or playlist or score our environments, internally and externally. In Essex, soundtracks are sand-tracks, ways across the silt, ways out from the shore and safely back in, following the signals (Fig. 7.5).

Soundscapes, like landscapes, are an environment. They are immersive. A soundscape can exist independently of any visual stimuli: close your eyes and *hear*. (Or, perhaps: close your eyes and *see*.) For me, growing up on a coastline, sand inevitably calls up the sea. I'm no desert-dweller. The sand is the doorway to the ocean, to another kind of immersion. If the sandy beach is an intertidal space, it is also the space of possibility, of becomings, of journeys. It is the space of transformation: of rock into stones, of stones into pebbles, of pebbles into sand. It is also the space of conjuration, of rituals: to go out, and to journey back. As in the cases of John Stonehouse or Reginald Perrin, you leave your clothes on the beach as a marker of your disappearance, a fraudulent dip into nothingness; or, as in "Everyday is like Sunday", you can come back from a swim to find your clothes have been stolen. A small stone taken from the beach is a memento but it is also something ritual. The sound of a shell, which when put to your ear seems to contain the sea itself, is a murmuration. The spiral of the seashell, the spiral of the cochlea, guides us inwards to the centre, and outwards to the edge. What we hear in the shell is the sea; what we hear in the shell is ourselves.

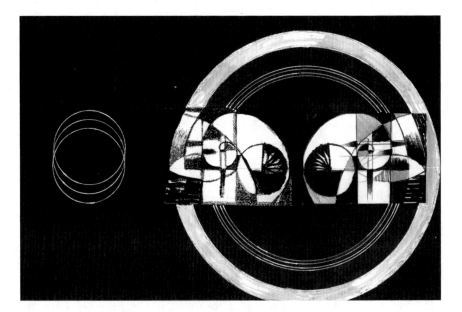

Fig. 7.5 *Soundscape*. Brian Baker, mixed media/collage, 2019

Notes

1. Robert Macfarlane, *The Old Ways* (London: Penguin, 2013), 69.
2. Macfarlane, *The Old Ways*, 67.
3. Macfarlane, *The Old Ways*, 73.
4. Macfarlane, *The Old Ways*, 80.
5. *Oil City Confidential*, dir. Julien Temple (London: Cadiz Music, 2009).
6. Eight Rounds Rapid, *Lossleader* (Cadiz Records, 2014).
7. Macfarlane, *The Old Ways*, 59.
8. Justin Hopper, *Public Record*, 2013, http://www.justin-hopper.com/public-record-estuary/.
9. The Who, *Quadrophenia* (Track Records, 1973).
10. Iris Murdoch, *The Sea the Sea* (London: Vintage, 1999), 1.
11. Cynan Jones, *Cove* (London: Granta, 2016), 1.
12. Jones, *Cove*, 10.
13. R. Murray Schafer, *Our Sonic Environment and the Soundscape: The Tuning of the World* (Rochester, VT: Destiny Books, 1994), 15.
14. Jack Kerouac, *Big Sur* (London: Penguin, 2012), 172.
15. Kerouac, *Big Sur*, 173.
16. Kerouac, *Big Sur*, 185.
17. Kerouac, *Big Sur*, 3.
18. The Triffids, *Born Sandy Devotional* (Hot Records, 1985).
19. Dante Alighieri, *Inferno* (New York: Oxford University Press, 1963), 147.
20. Dante, *Inferno*, 184.
21. Dante, *Inferno*, 171.
22. Dante, *Inferno*, 171–172.
23. Simon Critchley, *Notes on Suicide* (London: Fitzcarraldo Editions, 2015), 47.
24. Critchley, *Notes on Suicide*, 48–49.
25. Critchley, *Notes on Suicide*, 49.
26. David Foster Wallace, *This Is Water: Some Thoughts on a Significant Occasion, About Living a Compassionate Life* (Boston: Little, Brown, 2009).
27. Cocteau Twins, *Victorialand* (4AD, 1987).
28. Cocteau Twins and Harold Budd, *The Moon and the Melodies* (4AD, 1986).
29. This Mortal Coil, *It'll End in Tears* (4AD, 1984).
30. Morrissey, *Viva Hate* (HMV, 1988).
31. Schafer, *The Soundscape*, 9–10.

8

Rough and Smooth Sands: Social Thresholds and Seaside Style

Jo Carruthers

Mimicking liquid in its capacity to level out and take on the natural curvature of gentle undulations, sand's smooth surfaces have drawn writers and poets to think of it as psychologically absorbing and meditative. It was in the Victorian period that sandscapes became popular as respite destinations and novels of the time often feature the seaside as an early form of the health resort. Margaret in Elizabeth Gaskell's *North and South* (1854) visits the seaside first with her family and then by herself, both times processing traumatic life experiences on the sand. Near the beginning of the novel, during the upheaval of relocating from southern village to northern industrial city after her father's recent crisis of faith, Margaret finds a soothing "luxury of pensiveness" on the sandscape.[1] While Gaskell's novels are better known for their invocation of industrial or village life, *North and South* offers us a view of the sandscape as a kind of in-between space that enables respite from social and historical tangles.

On their voyage northwards, the Hales stop off at Heston, thirty miles from Milton, the Southport to Gaskell's fictional Manchester. At the seaside, Margaret first becomes aware of the differences of the north to her beloved south in what she sees: horse carts, people's busyness and a subdued greyness. The distinctions between north and south that are the heart of the novel testify to the fraught historical moment of industrialisation that many Victorians seemed to feel very acutely. The novel makes clear that Margaret, like

J. Carruthers (✉)
Lancaster University, Lancashire, UK

© The Author(s) 2020
J. Carruthers and N. Dakkak (eds.), *Sandscapes*,
https://doi.org/10.1007/978-3-030-44780-9_8

many of those at the time of Gaskell's writing, is living with loss, change and anxiety. It is only the sandscape itself that offers a reprieve from the weightiness of life. When on the sands, Margaret is able to live within a present moment unencumbered by past, future and social reality:

> There, for the first time for many days, did Margaret feel at rest. There was a dreaminess in the rest, too, which made it still more perfect and luxurious to repose in. The distant sea, lapping the sandy shore with measured sound; the nearer cries of the donkey-boys; the unusual scenes moving before her like pictures, which she cared not in her laziness to have fully explained before they passed away; the stroll down to the beach to breathe the sea-air, soft and warm on that sandy shore even to the end of November; the great long misty sea-line touching the tender-coloured sky; the white sail of a distant boat turning silver in some pale sunbeam:—it seemed as if she could dream her life away in such luxury of pensiveness, in which she made her present all in all, from not daring to think of the past, or wishing to contemplate the future.[2]

The sea's regular lapping suggests a soothing soundscape that is at first overlaid with unwanted cries and scenes. On the sandscape, boundaries and entities coalesce in the tactile softness and warmth of sea air, in the delicacy of the "tender sky", in the indistinctness of the misty sea-line, and in a sail transfigured, seemingly supernaturally, into silver by a sunbeam. Such haziness produces Margaret's welcome dream state and, through luxuriating in the sandscape, her present can transpose the past and future. Such sensuousness enables the distancing of cultural freight in a present overwhelmed by feeling.

Impelled by the desire to avoid past and future ("not daring to think"), Margaret welcomes this dream-like state which does not suggest a lack of thinking but merely a different kind of thinking, a positive sensuous and mindful engagement with and experience of her present surroundings. Here, repetition and indistinctness create an illusion of a sustained present. In a perpetual now, Margaret experiences a "perfect", "luxurious" and nurturing restfulness. While Margaret's seaside visit may be thought of as an escape, in line with Victorian understandings of the seaside as a place of healthy recuperation, it is not an escape *from*, but an escape *to*. Gaskell's sandscape is not a non-place or nowhere. Her sandscape consists of feelings: measured sound, softness, warmth, tenderness. The mutability and shifting nature of the sandscape and its environment becomes a resource of hope so that mutability offers a momentary reprieve from a socially and historically embroiled world.

Sandscape narratives often invoke this sense of looking out and losing the burden of the self. In a world in which urban landscapes and built environments overwhelm our senses with sounds, lights, flashings, signs, adverts and icons, here we find one of the few remaining landscapes that can seem to exist without human marks of any permanency. The sandscape's instability wards off human interference, to a certain extent at least. Where fields are full of pylons, mountains are marked by human-etched paths, stones, cairns and even cafés, the shifting and precarious sandscape is near impossible to build on and can be a place of anytime and anywhere. As observed from the edge of the sandscape, the sea beckons to our desire to be outside of ourselves, to be beyond the burdens of everyday life. The sandscape offers the alluring prospect of transcendence.

* * *

Walking along the sandscape in front of the iconic Midland Hotel in Morecambe (Fig. 8.1), I can appreciate Margaret's sand-and-sea-induced pensiveness. Looking out to sea, I too am drawn to the skyline, to the lulling movement of waves and clouds, to the circling and swooping of birds. Looking back, the contours of the white hotel are a soothing continuation of the shoreline. Its clean lines replicate the ribbon of sand, itself an echo of the horizon line of the sea. My eye moves easily along these contours, from a distant horizon along the shore, along the building's slight curvature, over to the range of Lake District mountains and back again to the skyline. All is soothing, seamless movement. It seems almost impossible to pay attention to any tangles in my mind.

The transcendence and "luxury of pensiveness" that elevates the individual beyond and above quotidian realities is precisely the aim of the straight lines of modernist architecture as represented in the Midland Hotel, designed by Oliver Hill, and opened in Morecambe in 1933. Hill built his hotel directly onto the sand—a surprising decision, perhaps, in view of the children's rhyme about the foolish man, and possibly regretted when remedial works were needed to mend structural cracks the year after the hotel opened. The hotel is three storeys high, with two wings that emerge from a central grand staircase and lobby. Interior and exterior alike nod to the hotel's sandy environment with the entrance laid with a buff terrazzo floor, replicating the sandscape.[3] The hotel's design expressed Hill's commitment to the complementarity of architectural and natural forms. In replicating the sweep of the coastline, it maximises the number of rooms with a sea view—following Morecambe Corporation's design for the new promenade—but it also makes the hotel a strange mix: it is a continuation of, at the same time as standing out from, its shoreline setting. In her chapter in this collection, Jenn Ashworth describes the hotel as being "like a clean white boat about to set off, or a spaceship, just landed"—not quite at home. Hill saw it as an opportunity to build "the

Fig. 8.1 *Midland Hotel*, Brian Baker

first really modern hotel in this country", with a style that valued health and nature as national treasures, as well as the simple lines of modernist aesthetics that indicated the futuristic promise of the early twentieth century.[4]

The Midland Hotel's minimal, clean lines invoke a sense of simplicity, amplifying the effortlessness of the sandscape that lies next to it. The interior is similarly sleek, smooth and modern, so that the magazine *Country Life* in 1933 could celebrate the fact that visitors "insensibly relax and feel peaceful".[5] In such places, life may be lived deliberately, and human nature might somehow be brought to its most basic and purest forms and desires. These simple lines invoke horizon, sea edge, gentle ripples of sand, cloud masses and streaks of sunlight that seem to enable a smoothing out of our very selves.

* * *

Simple things promise to us the meeting of inner and outer form—anything excessive or unneeded has been stripped away to reveal the thing itself. The literal meaning of simple is "single fold": *sem* means "one", and is drawn from the latin *simplex*. So the simple form is singular—leading to the association of simplicity with honesty. Simple things are not double, or duplicitous. Neither are they complex, a word again taken from the act of folding: here the folding together of multiple things (*com*—together; *plex*—fold or plait). Simplicity promises transparency—it is truthful, accessible and open to all.

Morecambe is famous for the Midland Hotel, but it is also full of non-conformist chapels and Protestant churches: there are even two Methodist churches a mere stone's throw from each other.[6] Walk a few streets back from the sands and a similar kind of pensiveness can be found in the white walls, rough-hewn pews and stark simplicity of a variety of dissenting chapels. Such chapels fit within a Protestantism that on principle rejects religious images, iconography and complicated rituals. These outward forms were dismissed as obstructions to a more straightforward faith and direct communication between oneself and God. Reformers in the sixteenth century were the original protestors for and advocates of simplicity, objecting against what they saw as the Pope's deliberately fangled authority in order to assert their belief in an unmediated—simple—relationship with the divine. This refusal of icons, images and ornament took the form of plain style, today so familiar in the simple lines and whiteness of these churches and chapels. Protestantism and modernism—these unlikely bedfellows—are similar in their suspicion of complexity, both averse to ornament and committed to the authentic.

Yet, the Midland Hotel and the nearby chapels don't embody the same *kind* of simplicity. In modernist styling, the simple became a mark of an elite taste. The very choice of minimalist style indicates discernment, and a taste for the elaborate therefore reveals a lack of discrimination. Just look at the makeover shows on TV that take someone with supposedly excessive style and put experts to work for hours to make them look "natural". Adherence to simplicity as inherently valuable leads to cultural and subcultural prejudice: those who don't value simple aesthetics are flawed in their judgement.[7] Newspaper feature writers occasionally turn against the fashion for simplicity gurus in order to question the current trend for minimalism and decluttering. In her opinion piece, Chelsea Fagan questions the push for simple living that promises a step away from consumer culture and the rat race. Fagan notes in passing that this is an expensive industry and available only to the well-off, ending her article with the scornful comment: "this kind of 'minimalism' is just another boring product that wealthy people can buy".[8]

* * *

The 1930s, when the Midland Hotel was built, was a time when many people held strong beliefs about the effect of architecture not only on the individual, but also on whole groups of people and even the nation itself. Architectural writings echoed the sentiment of the age which strived for national perfection and sought to design undesirable elements of society out of existence. Hill himself aspired to an architecture that would ameliorate those who dwelt in it. His embracing of functionalist architecture aligned with his commitment to build healthy environments to produce the settings for a more egalitarian society: his buildings prioritise open spaces and recreation. Hill was first influenced by Art Deco design through visits abroad: at the 1925 Paris Exposition des Arts Décoratifs and the Stockholm Exhibition of 1930, where he first encountered Scandinavian "International Style" or Functionalism as it is often known.[9] He disdained modernist architecture that demonstrated what he called a "lack of grace", perhaps invoking in this phrase the specifically Nordic "Swedish Grace", as exemplified in Stockholm Town Hall that presented a mixture of neoclassicism and Art Deco, just the kind of interior design chosen by Hill for the Midland.[10] However different, the romantic Swedish Grace and stark Functionalism shared an ideology to create, as Carl Markland and Peter Stadius note, "a certain type of mass society where the rate of class inclusion, and hence social levelling, would be greater than ever before".[11] However, it became clear that such levelling would not be without its casualties. It was a flattening of taste through elimination. By throwing out the crude tastes of the "lower" classes, it was believed that the people themselves would embrace a cleaner, better, more elegant and sophisticated, minimalist style and set of associated behaviours. Such simple and clean lines were believed in: they would produce simpler and cleaner people, pulling the masses out of their vulgar tastes that were often expressed by the use of too much colour, too much detail, too much pattern. The floor plan for the hotel, designed to keep staff and clientele separate, "was a clear diagram of a '30s class structure", Peter Davey asserts.[12]

The idea that modernist architecture had a moral function was its *raison d'etre*. It is telling that the Stockholm Exhibition included an exhibit by Herman Lundborg, the head of the Institute for Racial Biology in Uppsala. This scientist considered that "A populace material of good racial faculties is the highest asset of a country", a belief made evident in a discreet display but extended throughout the minimalist aesthetics promoted at the Exhibition.[13] Hill had already come across the correlation between ideology and architecture in realising a "healthy life" when he had visited Germany with the Architecture Association in 1929. There he engaged with the ideas of *Wohnkulture* ("Living culture"), which promoted healthy living as a direct influence on the political and social health of the nation. The biopolitical promise of architecture's contribution to a vigorous population is evident in Hill's own creed as expressed in 1931, two years before he starts building the Midland Hotel. "Directness, fitness, and economy", he writes, "are the paramount requisites, while simplicity, a maximum of sunlight and such facilities for recreation and exercise, dancing, swimming, and squash, as may be possible, will be large factors to be provided for in the modern house".[14]

Hill expresses a physical organicism—the belief that physical activity has a natural relationship to thinking and ethos—and links healthy living to the economic restraint of modernist architecture. The Midland Hotel fits the interwar national agenda on the state of the nation's health and management of its urban working classes.

* * *

When the Midland Hotel was built, it was the epitome of modernist, healthy design, whereas the town of Morecambe itself has been frequently berated for failing in its stubborn commitment to seaside gaudiness. In his *Designing the Seaside: Architecture, Society and Nature*, Fred Gray calls it the "out-of-place Midland Hotel", its clean lines unsuited to "Morecambe's decidedly Victorian coastal fringe".[15] Indeed, the hotel provides a space precisely for the smarter visitors to the city to drive to the hotel's car park, walk directly into the hotel, and walk from the rear of the hotel directly on to the beach. They could stay for days and drive off again without ever having to have stepped foot in the local area. Indeed, the modernist design of such buildings followed not an egalitarian agenda as expressed by Hill, but an elitist one. Filled with top-end cars, the car park faces the Midland on one side and cut-price stores on the other. Gray argues that the construction of such striking buildings was a deliberate strategy by municipal authorities to "segregate the fashionable and respectable from the unsavoury and rough".[16] Gatekeepers were employed to guard against what he calls "the invasion of the hoi polloi". The point, Gray argues, is precisely that such "out-of-place" hotels are meant to overawe. They are designed not merely to be calm, clean spaces, but to be "imposing": "too daunting for most working-class holidaymakers to venture across the threshold or risk the wrath of uniformed doormen".[17] Intimidating hotels like the Midland push away reticent visitors by separating the elegant, simple and sophisticated from the ornate, elaborate or crude.

The Midland Hotel is an icon to simplicity, and its elegant lines also shape the sandscape according to similar values. While I looked out at the hotel and the pale Morecambe sandscape as a place of escape, it seemed to matter what I am escaping from. And who can be included in that escape. I have to admit that I, too, am intimidated. I'm wearing my scruffy jeans, my cardigan has a hole in it and I'm wearing an unfashionable waterproof jacket that doesn't quite fit properly. I am not dressed for the shocking pink bar with its centre-piece crystal sculpture. In my mind, the hotel is always inhabited by people from its earliest generation: men in tailored suits and women in long white dresses, probably smoking from now unfashionable cigarette holders (the cigarettes not the holders) and sipping margaritas sparkling with salt. It may be that on sandscapes more than on any other land, we can believe we have escaped to a past or another kind of space that isn't tainted by social media, computers or the unsightly technology of a waterproof or hiking shoes. We may believe that this is a place that is cleaner, purer, brighter. That I too would also be bright and perfect if I were wearing an evening gown; that within those walls I won't have to meet anyone but the beautiful and the bright.

Modernist architecture conspires with the sandscape in the promise its smooth surfaces offer to elevate the elite above the full complexities of seaside living. This place looks like it is on perpetual holiday mode, but only for visitors like me and certainly not—I imagine—to the waiters, cleaners or chefs who work in the hotel. The simple architecture of the hotel is in cahoots with and yet also overwrites the idea of the sandscape as empty land. And this overwriting is also one of scrubbing out a more ornate Victorian seaside aesthetic that belongs to, and is celebrated and populated by, other people. On the opposite side of the road to the Midland Hotel is the Winter Gardens, a building—with its rust-coloured bricks and sage green ironwork—as grand as the hotel. Although usually closed, it can be seen in the 1960 film *The Entertainer* as the stage for Archie Rice.[18] It is used now only for occasional shows and exhibitions, although there are regular campaigns to try and restore it fully. To buy into the promise of escape and elevation that rejects the everyday also risks turning a blind eye.

* * *

In their book, *Modernism on Sea*, Lara Feigel and Alexandra Harris also note that modernist architecture is "out of place".[19] But being out of place is perhaps the point, and is where hotels like the Midland get their visual effect. Minimalism's aesthetic force comes from its contrast to ornate contexts, argues Edward Strickland and he illustrates his point with the example: "spare newspaper/magazine ads catch one's eye amid page after page of crammed columns of once-in-a-lifetime prices".[20] While modernist architecture is associated with calmness, there is some ambivalence in *Modernism on Sea* about the alternative it contrasts with: crude seaside aesthetics. Yet minimalism is not all good: calmness comes at the price of a certain coldness, with real pleasure located elsewhere. Feigel and Harris suggest that, "For the writer or filmmaker who was prepared to countenance the crude, the popular seaside resort offered the chance to develop an aesthetic of pleasure".[21] Pleasure is popular but the editors here, however much they celebrate its aesthetic, can't escape the tenacity of its association with the implicit negative values of crudeness. Feigel and Harris also identify this pleasure more specifically in the lavish and elaborate: "the popular seaside resort enabled an alternative artistic tradition that was opulent and crude but, above all, pleasurable".[22] Just as seaside play slips all too easily into the illicit, so pleasure seems to be conjoined with the undesirably crude.

In the Gurinder Chadha film, *Bhaji on the Beach*, when a young woman complains that her aunties, sitting in their coats on the beach, aren't having fun, one of them responds: "Well, when we start to enjoy ourselves people start to talk".[23] The aunties are constrained by social disapproval and so do not fully embrace seaside recklessness, although they do relax enough at one point to paddle in the sea. Having fun draws attention. The film narrates a trip made by a group of women of Indian descent to visit Blackpool. Perhaps because it focuses on cultural mobility, on the ingrained cultural habits of different generations and cultures, the film does a good job of representing seaside pleasure. As the "aunties" arrive, the camera pans over the various beach pleasures: donkey rides, a picnic blanket, multicoloured windbreaks, sitting on deckchairs in coats, playing ball games with a fuzzy yellow ball that sticks to Velcro pads, building a sandcastle in jeans. Later in the film we see a fairground, a Wurlitzer on the pier, a group eating fish and chips, scenes in a greasy spoon, and a Ladies' Only Night at "Manhattan's Bar and Dinette" featuring a stripper. The film's trajectory follows the logic of the writing in *Modernism on Sea*: the spectacles of pleasure are coarse, unpolished, maybe even crude or vulgar, but utterly enjoyed by these women.

* * *

The sleekness of sandscapes is a question of scale and perspective. As I stand on the seafront promenade—a ribbon of pavement that holds walkers, cyclists, buggies and scooters—the sand appears as strands, which is the title of Jean Sprackland's book on the Southport sandscape.[24] The sand is level and at times undulating. When I look in closer, these smooth lines become messier and less even. Standing on the sandscape and looking down at my feet, I can see not individual sand grains, but other alien material: grass, stones, a plastic pen lid. A movement catches my eye and I see a small insect—or is it even a tiny crab? It emerges for the briefest of moments to bury down again into the sand. Sitting on the sand, I can now see individual grains and even smaller vegetation and debris. I am reminded of the different objects that Sprackland finds in her daily walks on the Southport beach: the everyday discovery of a cup, the nearly supernatural unearthing of a long-buried prehistoric footprint captured on a re-emerging mud flat. I think of Elizabeth Bishop's poem about the sandpiper, this bird on the opposite side of the Atlantic, obsessively searching through grains of "quartz, rose and amethyst".[25] But of course, it was never the sand that the sandpiper was interested in, but the insects that nestle there.

The finickertiness of the sandpiper resembles the kind of scientific pedantry that led to the development of microscopes that were able to image sand grain surface topography in the 1960s.[26] The photographs of sand grains that went viral in the early twenty-first century introduce a seemingly alien landscape hitherto invisible to the human eye. They presented an imagined alien world in the lurid curves of green, pink and purple fragments of seashells. Of course, these are photographs of tropical sand. I suspect that if I saw the sand of Morecambe underneath an SEM microscope, made as it is from the cream and white shells native to the North West coast, there would be less colour. I imagine knobbly, cream coloured granules, looking more like enlarged lumps of brown sugar. Rough spheres.

The smooth surfaces of sandscapes belie a more complex substance. Sand is made up, as the book *Sand and Sandstone* explains, of a "framework of grains".[27] Sand epitomises the complex intertwining of substances that the anthropologist Tim Ingold encourages us to contemplate as he writes about the impossibility of a rigid boundary between earth and sky. We are unable, he argues, to identify the exact point at which sky becomes ground: "the wind can whip sand into dunes, snow into drifts, and the water of lakes and oceans into waves".[28] Strict boundaries are elusive. While the pebble beach may act as a larger-than-life image of the elemental merging between land and sky, the sandscape embodies those entangled boundaries that only appear smooth. Sand appears to produce a neat line between land and air and yet it exemplifies the blurred edge as air, water, stone, the biological, fauna, insects, and small animals intertwine. Sand is a messy business. Smooth from a distance only, sand up close is made up of a whole array of other materials.

*   *   *

The Victorians embraced the seaside holiday at the same time that they embraced the term "roughs" to refer to the especially threatening members of the lower classes. Parliamentary discussions and popular magazines debated the suitability of the "roughs" for the vote in the emancipation debates of the 1860s about the extension of the franchise beyond landowners. The worry was that the vote would be extended to the men of the lower classes who would not be educated. "Roughs" must have seemed a label that suited those who expressed liberal expressions of universal equality but also insisted on the unsuitability of those they considered not yet civilised enough to deserve the vote. Although it is an insult that has fallen out of use, people and places are still often described as a bit "rough", and seaside towns are the places that often seem to attract this epithet.

Seaside towns are undeniably some of the most hard-up areas of the UK. Michael Bracewell and Linder's celebratory and affectionate guide to Morecambe and Heysham was written in 2003 when the Midland Hotel was still derelict and their description of the seaside town sounds a melancholic lament over the lost heyday of seaside resorts more generally.[29] A 2017 report by the Social Market Foundation indicates that they have some of the highest unemployment rates and the lowest average earnings. Back in 1997, the Office for National Statistics showed that economic output per person was 23% lower in coastal communities than non-coastal communities; in 2015, the gap had grown to 26%. The chief economist and author of the SMF report Scott Corfe argued that part of the problem is "there is currently no official definition of a 'coastal community'".[30] It was only in 2013 that the ONS pooled together 57 of the largest seaside resorts, giving these places a shared identity and opening them up to statistical analyses.[31] These kinds of numbers are important indicators of the political neglect of the seaside resort and lack of employment opportunities, but they do not faithfully represent the particularities of any one seaside town—they do not tell us what it looks like or what it is really like to live there.

* * *

Seas are often described as rough, as is weather. Sand is put to use in rough sandpaper, even though it can be bought in grades from smooth to coarse. Sandscapes or sand are rarely described as "rough" in texture, even though sand itself, and the experience of standing on a sandy beach, or feeling grains on skin, is decidedly rough. Why are sandscapes and seaside towns often called "rough", as well as the people who live in them? Fred Gray mixes olfactory and tactile senses when he identifies those repelled by the Midland Hotel, discussed above, as "unsavoury and rough". The term "rough" became more popular in the Victorian period and novelists—Charles Dickens, Elizabeth Gaskell, Charlotte Brontë among them—explored the word and concept, quite critically, in their fiction. As early as *David Copperfield* (1849–1850), Dickens seems to have recognised the association between the rough and sandscapes. In this novel, those who dwell near or even on sand represent the lower class and familial acquaintances of David's life, whereas the city represents his more respectable, yet less satisfying, life. The novel pivots on the distinctions between the respected and polished well-to-do, epitomised in his friend Steerforth and his family, and the coarse and rough lower classes represented by David's nurse, Peggoty, and her family. David visits Peggoty at her family home in Yarmouth and there meets her brother, a Mr Peggoty, who repeatedly identifies himself as "rough". More specifically, he is "as rough as a Sea Porkypine", seaside dialect here expressing his distance from the fine articulation of his London visitors. Associated with a creature of the sand, he lives in a boat-turned-house that rises up from the sandscape.[32] What Mr Peggoty calls "rough" is experienced as a warm gruffness by David himself, an admirable loyalty and affection, that is preferable to Steerforth's gentlemanly and disingenuous smoothness. Peggoty's roughness, with his ungrammatical speech and ungentlemanly clothes, can only be comprehended in disparaging terms by the privileged and snobby Mrs Steerforth and her companion Mrs Dartle as "uneducated", "ignorant" and "common".[33] Dickens takes on the social connotations of roughness—and its association with the sandscape—to rethink on the page what it means to be rough.

Taking the Rough with the Smooth

The word rough invokes a specific kind of surface: unfinished, coarse, uneven. It is a word that is used in a surprising amount of phrases: rough justice, take the rough with the smooth, rough draft, feeling rough as a dog (or a badger's ass as my friends reminded me). It is an insult, certainly, but sometimes not: we talk about rough play—rough and tumble—being a bit "rough around the edges", even "in the rough". This last phrase is used of precious stones before being cut—hence the phrase rough diamond. It intimates an inherent value before the polishing activity of civilising processes, similar to the kind disparaged in *David Copperfield*. In terms of manners, the smooth and the polished suggest refinement but also a coldness and distance, whereas the unfinished surface of the rough suggests a more amenable interface that allows purchase.

Walking through the streets of Morecambe, I think about this town as "rough". Perhaps the label is an insult that cannot be re-evaluated, but thanks to its spectrum of meanings, it is never completely under anyone's control. In terms of being unfinished, Morecambe's roughness can be seen in the extent of vivid blue boarding covering up the now-defunct area that was once the amusement park "Frontierland". And there is plenty that was once finished and now in a state of disrepair: only half of the lights are working around the "Seagull Arcades" sign when I walk past. But there's also a pleasing mish-mash of shops, quite different to the polish of the high street line of chain shopfronts. There is a vintage café, "The View", and "The Old Pier Bookshop", a warren of over-packed shelving. In the

West End, I come across a row of terraces on West Street that have a beautiful Victorian wrought-iron awning. In these buildings can be found The Exchange, a small community arts venture. With appealing yet higgledy-piggledy shelves—rough, one might say—the shop sells locally produced work and has a workroom attached in which the women sitting there—all involved in some kind of craft-making—invite me to join them and tell me about different events going on across the town. When I come back to visit months later, the shop is closed but there is a piece of weaving outside left in mid-construction, coloured ribbons on the cream and burgundy iron railings, a polka dot rainbow-coloured toy windmill in a flower pot, and wind chimes hanging from the canopy. It is raining and I am grateful for the protection as I look at the playful ornaments.

On this return visit, I head to the Old Pier bookshop to see if I can buy a book on Morecambe and find R. C. Quick's 1963 guide to the town, *The History of Morecambe and Heysham*. I have not got any cash, so I have to walk through the rain to the hole in the wall. I walk past a Costa, and a closed-down "New Look Fashions" in which joiners are working behind the shop-front boarding. On the way back to the shop, I see Pedder Sweet News. A poster outside boasts that it has been there since 1903. I go in and feel a thrill of memory—nostalgia you might say—at the sight of row after row of sweets in tall, rectangular jars. Hand-written labels hold bon-bons, white chocolate mice, wine gums. I pick up a couple of sticks of "Morecambe rock" striped with rainbow colours for my children. I am meant to be getting back, but I also pop into "Pottyroo's", a paint-your-own-pottery shop to find out their opening hours. I plan to take my children there, but wonder if I will be able to drag them past the flashing lights of the three amusement arcades I see. I make a mental note to stock up on small change before I return.

Notes

1. Elizabeth Gaskell, *North and South*, ed. Alan Shelston (London: Norton, 2005), ch. 7.
2. Gaskell, *North and South*, ch. 7.
3. Peter Davey, "Morecambe and Modernity: Union North and Avanti Architects Help Revive Morecambe's Midland Hotel for Developer/Hotelier, Urban Splash", *Architectural Review* 224, no. 1339 (September 2008): 76–81 (78).
4. Barry Guise and Pam Brook (eds.), *The Midland Hotel: Morecambe's White Hope*, 3rd edn. (Lancaster: Palatine Books, 2009, repr. 2012), 15.
5. "The LMS as Maecenas", *Country Life*, LXXIV (18 November 1933): 539–544, cited in Davey, "Morecambe and Modernity", 80.

6. R. C. Quick, *The History of Morecambe and Heysham* (printed by *Morecambe Times*, Queen Street Morecambe [1963]).
7. For a cultural history of simplicity and Englishness, see Jo Carruthers, *England's Secular Scripture: Islamophobia and the Protestant Aesthetic* (London: Continuum, 2011).
8. Chelsea Fagan, "Minimalism: Another Boring Product Wealthy People Can Buy", *Guardian* (4 March 2017).
9. See Davey, "Morecambe and Modernity", 78; Guise and Brook, *The Midland Hotel*, 20.
10. Colin St John Wilson, "The Fields of Her Former Triumph", *Review of the Stockholm Exhibition 1930: Modernisms' Breakthrough in Swedish Architecture*, by Eva Rudberg (Stockholm: Stockholmia Förlag, 1998).
11. Carl Marklund and Peter Stadius, "Acceptance and Conformity: Merging Modernity with Nationalism in the Stockholm Exhibition in 1930", *Culture Unbound: Journal of Current Cultural Research*, 2 (December 2010): 609–634 (616).
12. Davey, "Morecambe and Modernity", 77.
13. Quoted in Marklund and Stadius, "Acceptance and Conformity", 623.
14. Oliver Hill, "The Modern Movement", *Architectural Design and Construction*, 1 (September 1931): 461–463 (461).
15. Fred Gray, *Designing the Seaside: Architecture, Society and Nature* (London: Reaktion Books, 2006), 52.
16. Gray, *Designing the Seaside*, 53.
17. Gray, *Designing the Seaside*, 54.
18. Tony Richardson (dir.), *The Entertainer* (London: Woodfall Film Productions, 1960).
19. Lara Feigel and Alexandra Harris, "Introduction", in *Modernism on Sea: Art and Culture at the British Seaside*, eds. Lara Feigel and Alexandra Harris (Witney: Peter Lang, 2009), 2.
20. Edward Strickland, *Minimalism—Origins* (Bloomington, IN: Indiana University Press, 1993), 2.
21. Feigel and Harris, "Introduction", 16.
22. Feigel and Harris, "Introduction", 17.
23. Gurinder Chadha (dir.), *Bhaji on the Beach* (London: Channel Four Films, 1993).
24. Jean Sprackland, *Strands: A Year of Discoveries on the Beach* (London: Jonathan Cape, 2012).
25. Elizabeth Bishop, "The Sandpiper", *Poems* (London: Chatto and Windus, 2011), 129.
26. William C. Mahaney, *Atlas of Sand Grain Surface Textures and Applications* (Oxford University Press, 2002), 7.
27. F. J. Pettijohn, Paul Edwin Potter, and Raymond Siever, *Sand and Sandstone* (New York: Springer Science, 1987), 1.

28. Tim Ingold, "Earth, Sky, Wind, and Weather", *Journal of the Royal Anthropological Institute* 13 (April 2007): S19–S38 (S32).
29. Michael Bracewell and Linder, *I Know Where I'm Going: A Guide to Morecambe and Heysham* (London: Book Works, 2003).
30. "Seaside Towns Among Most Deprived Communities in UK", *Guardian* (4 September 2017), https://www.theguardian.com/inequality/2017/sep/04/seaside-towns-among-most-deprived-communities-in-uk/.
31. "Deprivation in Seaside Towns", *Poverty and Social Exclusion* (20 August 2013), http://www.poverty.ac.uk/editorial/deprivation-seaside-towns.
32. Charles Dickens, *David Copperfield*, ed. Jerome H. Buckley (London: Norton, 1989), ch. 21.
33. Dickens, *David Copperfield*, ch. 32.

9

A Morecambe Mystery

Angela Piccini

Fennel for good. Sorghum for bad. Seeds of Syria on the Morecambe Sands.[1]

They call me La Tabacca. My business? Providing wicca-nomics solutions for European seaside towns that have seen better days. Towns hell-bent on blaming strangers for their failed fortunes. Towns with a dark heart but a lot to offer. I ride the black cat and cast sorghum and fennel spells to solve mysteries and generate prosperities.

Jacopo Panfilo rang me about the case. They call him The Patriarch but I'm having none of that. Ursula says it's because he's the most tenacious litigation lawyer in town. I say "smash the patriarchy".

I meet Panfilo late on the 18th. He's a partner at Baines Bagguley Penhale, the solicitors in that austere little stone number on the Bay. He suggests we go

A. Piccini (✉)
University of Bristol, Bristol, UK

© The Author(s) 2020
J. Carruthers and N. Dakkak (eds.), *Sandscapes*,
https://doi.org/10.1007/978-3-030-44780-9_9

to the little café on the promenade, overlooking the beach. We order coffees as the staff clear up around us, getting ready to close. Panfilo nods towards the water where a couple sit, head-to-head, deep in conversation and chips. Panfilo tells me they're talking about what happened to Panzona, Demdike's wife. What happened at the Midland all that time ago. Everybody's got a theory. Everybody's heard a version of the story. But what's certain is that Panzona's dead and I've been brought here to unravel the knots and lift the curse.

What's the lowdown, Panfilo? I'd read the news in an old copy of *The Westmoreland Gazette*, so I know at least some of the facts. Local woman found dead in Josaphat Field, some thirty miles away, near Pendle. The weird thing was, there were no obvious signs of violence. She'd been found fully clothed, submerged in one of the ponds up on the blanket bog. Weighed down by stones. Even stranger, a single strip of fabric had been removed from the back of her shirt, to reveal her spine. Panfilo tells me that Panzona Demdike had been a tearaway in Morecambe's Northern Soul scene, back in the 1980s, when she was sixteen and fuelled by Pro-Plus and endless cans of Quatro. She ran with the Pendle Witches, who were prone to pissing in people's milk bottles in the early hours of the morning. She grew up a bit and more recently had had a hand in bringing Northern Soul back to the town and in doing up the old places. These days she preferred to dance in the Rotunda Bar at the Midland after a couple of their gin and tonics. All poshed up with juniper and samphire. But the huge Christmas parties at Morecambe Winter Gardens gave her an excuse to really shine. To shake out her old Fred Perrys and Roxy Threads. While some of her friends had disappeared into the anonymity and comfort of George by Asda and a generous bit of padding, Panzona refused to be aged and softened.

After a lifetime filled with unremarkable events like jobs, marriages, kids, dissatisfactions and a bit of the grinding is-that-all-there-is-ness, Panzona got political. Started volunteering with One Woman at a Time after she heard what was happening to those women and girls who'd been settling in town.

While she had never left the northwest, she was curious about the world and felt a hot anger whenever she heard about injustices, particularly the squashing of women's rights. And while some of her neighbours muttered angrily about immigration, Panzona had sharp eyes and knew the problem in her town was down to decades of under-investment. All you had to do was watch those videos from the mid-1980s on YouTube to see what Thatcher and everyone since had done to this place. But by now, Panzona was mainly angry at the way that people's differences were being exploited by everyone, with the girls suffering the most.

Of course, there was a back story to this fierce interest of hers, which I had a hunch about, but I wanted Panfilo's version. I had no idea, then, where this was all going to lead.

Panzona's—how shall I say it—activism? She was just a kid really. That dangerous, magickal cusping age. When you think you're an adult, filled with desire, ambition, dreams and a beyond-your-own-years sense of responsibility. She was looking after her little brothers and sisters, a few cousins and some of the neighbours' kids too. Their parents all ran the rickety little bed and breakfasts in the West End. That is, until the tourists ran out. The kiddies loved it. Endless sausage and chips and almost no adult supervision because they were too busy stripping beds, pouring pints, massaging the accounts and sometimes the visitors, too. Panzona would take the little ones down to the pier to doss around. There was hardly anything to do in Morecambe by 1980. The Super Swimming Stadium shut in 1976, followed by the Winter Gardens and all the cinemas. West End pier—which they'd considered their own personal playground—had just been demolished after the 1977 storm. They didn't have the money for Marine Land or the Amusement Park. But there was open-air roller skating on the Central pier. Sure, that closed in '86. But before that, Panzona and the kids could sweet talk Lily Webber into letting them on the rink for free. Kept them out of trouble for a while, though of course when Panzona was a bit older she was sneaking into the all-nighters at the disco.

They arrived on one of those greyish, warmish days that failed to entice the tourists away from Tenerife to Morecambe. Menichino and Bosco del Merlo rolled into town in a beige Ford Granada. No one noticed them. Except me. They stayed at Panzona's folks' B&B Twin room, so they didn't stand out as spivs or anything. Panzona told me that they stayed for a few weeks. They said they'd come to town to set up a used car dealership. They were quiet, friendly. They liked to ask Panzona questions about school and hobbies. She liked the way they showed an interest in her. It seemed that to them she was more than just a babysitter. They spoke to her like she was a grown-up. So, when they invited her to go to the picture house, her parents didn't really think twice. At least, not at first.

The third time they suggested the three of them go to the cinema, Panzona's mum and dad caught something in the looks of the men and in Panzona's mood that made their hearts skip a beat. That night, Panzona's father invited me, Menichino and Bosco del Merlo for a nightcap. I was just a trainee solicitor at the time but had helped out the local hoteliers when organised crime started to make inroads in the town. Maybe you would call my connections a kind of protection racket but I did it because I had my own fish to fry and sometimes you have to be pragmatic to get a job done, right? Anyway, Panzona's father took a dusty bottle of Glenfiddich from behind the bar and poured us all doubles. The fact that I was there should have been warning enough for them. But they didn't bat an eyelid. Not much was said, but they knew what was behind the drinks. What I recall is that Panzona's father took a book down off the shelf. *The Night Battles*. By some guy called Carlo Ginzburg. I'd never heard of it. But he opened it and asked the men if they'd leave Panzona alone if he gave them "two bushels of wheat and two jars of wine". I couldn't figure that one out but they agreed and they stopped taking Panzona to the movies.

Menichino and Bosco del Merlo's car dealership grew and I didn't think much about them in the years that followed. Until the mid-1990s, when their rivals Enzo Marconi and Peter Grantham were found murdered. Terence Clifton went down for it, but I had my suspicions.

Years later, not long before her death, Panzona told me a little about that time. Instead of leaving her alone, they were constantly around her, at the school gates, at the roller rink, on the streets saying "if you had come you would have seen so many beautiful things". She told me with a shrug, "I was just a young girl and foolish, and I did go". Panzona seemed to think that Menichino and Bosco del Merlo were involved in the cockle picking gangs and, more recently, in providing cover for the men behind what was happening to the young women who were moving into town. I said she was crazy and paranoid.

I've been wracked with guilt since. But that's no good to anyone now. They're still out there. Doing whatever it is that they do. With impunity. I've tried and tried to get some dirt on them but nothing sticks. That's why I asked you here.

I thank Panfilo for sharing. Jesus. Men. The one guy who coulda made a difference failed that kid. She might still be with us if people had only thought for a minute that this doesn't have to be the way things play out.

#NotAllMen they say. Panfilo's clearly suffering through it all but so what? I'm angry for Panzona. I'm angry for this town.

On the way back to the hotel, the enormity of this job hits me. It sucks my heart towards that black place like these sands suck at my feet. A kid on a bike pops a death-defying wheelie. Just as I'm thinking about complimenting his skills, he shouts back at me—"Hey, you're La Tabacca! You suck blood from little kids. You make us sick, stunted. Die". News travels fast but the old witch lingo dies slow. It's not like I got into this business not knowing the bad rap we witches get. But hearing this at the tail end of Panzona's story is almost too much to bear.

I drag myself back to the Midland Hotel. It's ironic seeing Eric Gill's work in this setting. Particularly Triton and Neptune raunchily splashing around with a couple of under-age mermaids over the staircase. I'm in no mood for this. I head to my room for a bath and a lie-down. This is where I do the real work. Panfilo clearly wanted this thing solved. He booked me into one of the rooms overlooking the sea. I open the windows, breathe in the iodine and watch the tide rush in hungrily, swirling in eddies around the endless ridges of mud. I drop handfuls of sorghum and fennel seed into the tub. A cup of milk in the water, a glass of wine in my hand. "*Benandante e malandante, volesais balâ cun me?*" "Good witches, bad witches, will you dance with me", I say in Friulian. Times like these I gotta bring the old ways into play and, anyway, Morecambe and Lignano are connected by more than just sand. Funny how these resorts seem to blossom in witch-scapes. I lay myself out on the bed and let the breeze bring goosebumps to my skin.

I fall into my trance and Tagliamento, the black cat, jumps up onto the windowsill. I jump onto his back. I'm ready.

We prowl, unseen. Through the empty hallways, up the stairs to the top floor and out onto the roof. It's getting dark but we're alone up here. We head back down the spiral stairs to the lobby and pad across the seahorse mosaic, out through the grand entrance. Outside the hotel, it's dead quiet. White vans occasionally drive into the car park, unload, reload and exit again. For an under-occupied hotel, there's a lot going on behind the scenes. We sniff around the door to the kitchen. The menu seems good. Tagliamento and I fancy a bit of the oxtail that's bubbling away.

Another van pulls into the hotel and drives up to the kitchen entrance. Two guys get out. They're both non-descript tough-and-buff. Dead eyes, nicely shaped eyebrows, overly smooth arms and smelling of manufactured pine forest. Sort of *espace quelconque* men. They remind me of that Sam Beckett line—"Neither here nor there where all the footsteps ever fell can never fare nearer to anywhere nor from anywhere further away". The men open the back of the Berlingo, pull out what looks like a laundry trolley and roll it into the kitchen. They return with a similar trolley. It's obviously lighter than the one they just delivered and they sling it into the back of the van and drive away.

Tagliamento and I approach the laundry trolleys in the back of the kitchen, near the goods lift. The trolleys smell of fear, young women and opiates. In our trance state, we can't do anything about the everyday world right now. We need to know what's going on.

Back outside. The wind has died down and the tide is nearly in. The pervasive smell of mud is replaced by salty-sweet water. We hear a noise from the roof. Looking up, we're just in time to see Bosco del Merlo spin round on his heels and head back inside. That posture is unmistakable. It says—not a care in the world. But he was looking at something. For something? What's he doing here? Just then, a movement out of the corner of my eye catches my attention. A curtain twitches in the top floor corner room. And then I see him. Menichino. His eyes seemingly burning through me though I know that can't possibly be the case. He may be the Devil himself, but my magick is stronger than his.

Me and the cat turn around to face the sea and scan the horizon for whatever it might be that's worrying the two men. I can see a light flashing across the bay, somewhere near Humphrey Head. Must be that Haven holiday place. A nobby, one of those old shrimping boats, sails into the Green Street breakwater. It's too far away to see properly, even for Tagliamento, but we can just make out the silhouettes of two men, with a smaller figure held between them. Now I know what I need to do.

Back in my room. Tagliamento has melted into the night and I'm recovering from the agonising disorientation that the trance-journey causes. I look out from what is Bosco del Merlo and Menichino's point-of-view, over the bay. The flashing across the water has stopped. A man walks past the old electricity substation, down from the stone jetty and back towards the breakwater. He walks just like one of the men who's come off the nobby. Maybe I'm putting six and six together and getting the number of the beast, but it seems pretty clear to me that all of these things are connected.

Follow the money. Gangster-entrepreneurs, innovating on the fly as they navigate changing laws, shifting populations, altered fortunes. The people they exploit are resources, pure and simple. Why would they care given no one else seems to? My job is to be hard-nosed about it. To put a stop to the criminality, to seek justice, and to work magick. To craft a more compelling narrative, a spell for these towns and the people in them. I call it wicca-nomics, while down in the southern cross, those other witches, Gibson-Graham, called this stuff "diverse economies". But even this has been co-opted by the suits. I hear tell of a new kid on the block. "Inclusive growth" is now on the lips of every Local Enterprise Partnership and even the heritage people are after ways to monetise the past for the benefit of all. One-legged, Pendle witches with brown babies forming their own blood-shit-hair plastering companies for the National Trust? If you can spreadsheet it, it's a success! I'm not really that cynical, though. These traditional lands need something that isn't just nostalgia for dying white people. If I can only nail these guys before tomorrow, maybe there will be a chance for these girls and the ones who come after them. A chance to bring a new spirit into play.

I get dressed and head down to the Rotunda Bar for that drink. I sit near the bartender, with my back to the door. Soon Menichino and Bosco del Merlo show up, with a group of young women whose outfits don't do much to hide their ages. Although there's no way they can know who I am, I feel sick-scared. The atmosphere is tense and shot through with forced "fun". The girls laugh while the men order them drinks. It reminds me of that time in Tenby when I spent an evening with old alcoholics at the Lifeboat Tavern. That was a gig involving gangs dealing Ecstasy and "cut-and-shut" car breaking yards. Back in the day. These old guys seemed harmless at first, but that same edge was in the air. I'd learned how to manage hidden male aggression with my own dad so I thought "smile sweetly, don't challenge, feed the ego, watch out for traps". It's what these kids were doing at the bar. And it's a dance, because these guys know they're doing it. They get off on it.

A few drinks later, the group gets noisier. More men arrive. I recognise faces from the paper—councillors, cops, the head of the local business group.

The bartender pretends not to notice any of it. I'm middle-aged invisible so I can carry on sipping my grappa. Around ten, there's an unspoken shift in the room. The girls pick up little phials of citrus oil from the bar, dab it on their wrists and head upstairs to dance with the men.

I act quickly, head outside and call Panfilo. I ask him whether there are any not-bent police in town and how quickly they can get here. Panfilo tells me to leave it with him and hangs up. Within five minutes two unmarked cars turn up and a journalist friend of Panfilo's, there to document what goes down in case of future shenanigans. I ring up Panzona's co-worker at One Woman at a Time and tell her to get some support down here as soon as she can.

The sun rises over the Pendle hills. The tide is a million miles away and the sand looks dry enough to stand on without risk of sinking into it up to my belly. It's time for the ritual to begin. I imagine the girls we pulled out of the hotel last night. I imagine the evidence Panfilo now has to put those men away. He rang early this morning to say that they'd come clean about Panzona, too. How she'd found out what they'd been doing to all these girls and how she went to the police, but she'd chosen the wrong one to trust. They killed her like that because they are *malandante*. Bad witches whose bad magick draws on the chthonic power that sits at the bottom of the watery bogs up on the hills. And they leave the spines of their victims exposed so that the other night creatures can take their pleasures there.

I wish I'd known Panzona. I imagine her face lit up with joy on the dance floor, her righteous anger, and the work that she did in this town. I imagine the still-empty shops and the derelict bed and breakfasts, now bought up by property developers. I imagine all the ways in which this cold, muddy place has been seen as a place to be seen, a place of desire and fun and possibility. I imagine what that could look like now. My job is to conjure the glimmering outlines of a not-quite-seen. A starting point, not a goal. What they do with that will be up to them but I am here to hold it, to breathe softly into the embers of an idea. To find a way to make it ok to not know but to try anyway.

But what about Morecambe's traditional economy? How do all these things fit together? Google tells me that the value of Morecambe's goods and services has grown by more than £21 million over the past few years. Manufacturing and construction are the things and this place is gunning to be a powerhouse in the north-west. James Cropper pulp and paper, Gilkes and Marl International and Oxley Group LED technologies, Anord-Mardix cloud computing and Tritech and Siemens Subsea underwater video and sensing technologies. "Clean industries" that rely on the dirt being somewhere else. Lousy gender representation too and the pay gap needs a ton of work. Together it's like Morecambe is being transformed into a James Bond plot. And I imagine what it would be for these girls and women to work here, to determine the flow of data to help with the processing of fat video from beneath the thick sea, where LEDs illuminate the cockle beds. If women were 50% of the workforce rather than 2%, what could happen in this place? Like I said, a starting point.

I trace a wide circle. "Fennel for good. Sorghum for bad. Seeds of Syria on the Morecambe Sands. I mark you against witch, warlock, *belandante* and *malandante*, that they may neither speak nor act until they have counted the threads in the linen, the needles on a thorn-bush and the waves in the sea, that they may have nothing to say or do about you nor about any other woman".

Acknowledgements With thanks to and apologies for creative licence in using: Gibson-Graham, J. K. "Diverse Economies: Performative Practices for 'Other Worlds'." *Progress in Human Geography*, 32, no. 5 (2008): 613–632. https://doi.org/10.1177/0309132508090821. Ginzburg, Carlo. *The Night Battles: Witchcraft and Agrarian Cults in the Sixteenth and Seventeenth Centuries*. London: Routledge & Kegan Paul, 1983 [1966].

Note

1. The inspiration for this story comes from Carlo Ginzburg's *The Night Battles: Witchcraft and Agrarian Cults in the Sixteenth and Seventeenth Centuries* (London: Routledge & Kegan Paul, 1983 [1966]). The book is a microhistory and translation of Inquisitor records from northern Italy focusing on the cults centred around Latisana and the river Tagliamento in Friuli, northern Italy. Several generations of my family, including my father, are from that area. The Tagliamento flows down from the foothills of the Alps and empties at the seaside resort of Lignano. I was struck by the geographical, historical and contemporary political similarities between the Morecambe landscape and my own heritage "sandscape".

10

Map of the Quick

Shona Legaspi

It was an idea at first. Hastily scribbled lines on a piece of cloth because when the men died, the women outlived them by a Yorkshire mile. Annie was the first from her particular neck of the woods. She took the first steps and, after, others could follow. Before a journey, the cloth was copied and each time embellished, adorned with keepsakes that marked conflicted affections and difficult decisions. Charcoal scribbled lines became stitched lines and the cloth was passed between women's fingers, held up for inspection amid dimmed lights in damp filled darkened rooms, in rows of tightly packed terraces. The cloth was stuffed into the pockets of women who made their way to guest houses in Morecambe, where it was kept by some; let go by others as it fell, often unseen, onto sand. The map of generations wrapped in tight knots of cotton and wool pulled fast by ancestors; lines drawn across the land east to west, Yorkshire to Morecambe, by the human waste products of industrialisation.

We move
between
land and sea
land and sand

We draw our own lines
Left to right
We make our own maps.

S. Legaspi (✉)
Lancaster University, Lancashire, UK

© The Author(s) 2020
J. Carruthers and N. Dakkak (eds.), *Sandscapes*,
https://doi.org/10.1007/978-3-030-44780-9_10

My grandmother was a seaside landlady and she had been doing that job for quite a long time as had Annie, her mother before her. My great grandfather you see died working in Yorkshire. They had thirteen children so she decided to come to Morecambe to try to run a guest house. Both women were seaside landladies doing a business on their own.[1]

Annie
(arrived in Morecambe 1852)

What do you do with thirteen children and a dead husband? Where do you even start? The factory and the workhouse beckoned, for all of them, so Annie gathered up her wares and her children and she waded in, went fishing along the coast of Lancashire for a home, one they could all turn their hand to. She parted the dark, temporarily, and made a decision to walk away from the home her children were born in and the factory that had, until George's death, dominated and sustained her whole family. They had both worked at the worsted mill but Annie hadn't been there at all since the accident. The decision to leave had come quietly in the first few days. It sat so unobtrusively that she didn't even notice. Only the neighbours thought it remarkable but Annie recognised a friend as soon as one sat with her. She left the three eldest behind knowing they would willingly come to visit; gladly come to stand and marvel between the sky and the sea, a seamless pallet of endless blues where the light infiltrated every pore and warmed every bone. If the truth be known, Annie had only been to Morecambe once before, but sometimes, when life hits you full in the face, that's all you need to make the jump.

They made their way to Morecambe, a calling you might say. Annie was pushed to do it because she couldn't stay helpless and watch anymore die. She'd done that with their third born and now her husband. And she knew some of the others weren't long for staying, so she took the chance to move and was one of the first. She watched as other women with children followed east to west when they got the chance. Over time and generations, they migrated towards the light and Annie would watch them, on the sand, where many things are placed and held.

Annie didn't have a plan, not exactly, but she knew where to go and what she needed to do. She quickly found an arrangement for herself and her eldest daughter, Rosie, in service at the same grand house: two rooms for all of them. The landlord, Mr Thackrey, Yorkshire born and bred himself, was a property man, soon to retire and spend his days on the West Coast over-looking the Bay. With so many comings and goings in the town, he needed someone he could trust and agreed, at her suggestion, to let Annie and her family look after his houses until they were occupied. Annie was a forthright woman and scared of no one, Mr Thackrey could see that. She had disarmed him and, self-made himself, he could afford to be curious. Or perhaps it was compassion. Whatever the source, he assured himself, it was certainly not contrition. He had made his money through straight up hard work and good old fashioned "nowse". Annie understood the opportunity and she proposed to buy one of the properties, over time, a payment every few months. Mr Thackrey didn't think it possible but she made a promise and he gave her the benefit that his ample funds, amassed through the wool trade, allowed. In this way, Annie became the gratified owner of a guest house, open for the season and able to accommodate her entire family. Mr Thackrey did not regret the generosity he had bestowed upon her and Annie took up what she regarded as her rightful place: on the front, watching the Irish sea come and go, her attentive eye trained on the sand and all the things people left there.

* * *

Sooted and salted they came, wearily across the Pennines with their contam-inated blood and their deep contagion. They spilled, loaded and hacking across the land as they left their mark, the route etched in black and red. Annie watched; she always watched so she was always the first to know what was coming. She lived from her instincts and guided her family by them. Sometimes she didn't even need to watch, she just knew. Things came to her and she trusted that she knew.

As they travelled, Annie felt drawn to watch the creatures whose flight of freedom her children envied and she saw the families' movement west mirrored in flight throughout the sky. She saw all manner of birds from her train carriage and noted how they changed as the train edged towards More-cambe: warblers, linnets and paddocks became seafarers and waders that told her their destination was near. Annie was brought closer to an understanding that, like birds, humans also take flight when threatened or pushed to the limits, when the only solution is to up sticks and not look back. At home, Annie had seen her situation as clear as if she were a peregrine looking down on herself, clear, wise and sharp eyed. Birds were her messengers and despite the factories upending most of the landscape when they came, birds still appeared to Annie. There had been so much chatter among the women of hope and good things with the coming of the mills; the huge buildings, the hypnotic rhythm of machines that didn't stop working, but Annie had come to think that nought good had come of it for most folks. And now she was leaving everything she knew behind, trusting to God and her own instincts.

Morecambe

Coloured by grief, their wide eyes opened on to Morecambe Bay. This Bay that plays host to the sweep of the Irish sea which, twice daily, the moon pulls back and forth to reveal a vast expanse of sand as far as the eye can see. For centuries, many things, unwanted or precious things, had been buried deep beneath the sands. It too remained a burial ground for unfortunate souls caught by quicksand or a swift incoming tide that swallows everything it approaches. Despite the caution, or maybe because of it, as anyone fortunate enough to experience the Bay could testify, it is astounding in the intense beauty of its landscape. The Bay opens and speaks to the soul. A formidable setting and one to be wary of (Fig. 10.1).

Annie's family were courted by the seaside. They arrived in Morecambe carrying what little they had in boxes and bags, ten of them, the oldest holding the youngest, quietly taking it all in. It wasn't quite the season so Morecambe was only sparsely populated when Annie and her family landed, at the front, part of that first line of settlement that stood before the quick of the Bay, an endless borderline and their spitting basin of uncongested sand. Off-comers the locals called them even though they didn't come that way, off the sand, but it was the first place they went to. The sand met them and enacted a welcome the family did not expect. The children hesitated, their

Fig. 10.1 The open skies of Morecambe Bay © Shona Legaspi

surprise getting the better of them before hastening onto the sand. They had never seen anything quite like it and were immediately taken by the vast and surprising sandscape. The family poured so much of themselves into the sands; they had to stay. But always at the margins, in humility, a nod to the unequal order of things; the things they couldn't quite shake off even in their bid to do so. It wasn't far from one side to the other, Yorkshire to Morecambe, but when they arrived and stood in between the vast expanse of sky and sand, dumbstruck by both, they knew why they had come. They came here to escape, and it was a whole world away, all ninety miles of it.

There were no words for the surprise waiting for them, only the silent drawing in of chests, as the joy overtook and overwhelmed them. It gripped and pinned itself to their bodies, a sense of relief and rising reassurance that moved Annie to tears. Held in the eternal openness and enormity of the sky, something invisible and weightless that touched everything. All they needed to do was stand underneath, embraced by the light and the sun, the huge endless luminescence above the sea. It came uninvited and everyone felt it, even the youngest who stood, her face upturned towards the light, arms high, trying to touch the sky. There were days when they couldn't avoid but talk about it. Everyone had to say something. Their stunned and inadequate words trying to articulate something intangible, something godly the light brought them to. They knew, the family discussed it and they knew: it was love, a flooding gesture that filled them and didn't ask anything in return. It just came and on those days they were all joy and wonder. It topped up their reserves and imbued them with the kind of exuberance they had not known before.

It wasn't long before the railways started bringing whole factory loads of visitors.

"The labouring classes, and common inebriates", Thomas, the eldest declared over breakfast as he repeated to them what he'd heard.

"Stop listening to folks who don't know what they're talking about", Annie reprimanded, watching him closely. "They're just like us and they're welcome to all the courtesy we'll give them".

In the few years they had been there, Thomas had surely taken on airs and graces that were not his own. Annie had caught him trying on a new voice, mimicking the gentlemen folk he'd come across on the odd job he'd done at the Midland hotel. She'd laughed but she could see he was pained by how the posh folks treated him. She could see she would soon lose him to other people's ambitions.

Breakfast was the main time they all came together, except for the youngest who stayed sleeping. They were up before the guests during the season and

Annie dished out the extra jobs on top of what they all had to do each day. Washing days were the worst, wrapping themselves up in bed sheets was the only way the children could carry them. But the stink was overpowering sometimes so they would throw them down the stairs when no one was looking, then take them out the back to be doused and scrubbed as quick as they could. They were long days. Their guests were conditioned to getting up at four in the morning and no matter how they tried, it didn't change just because they were on holiday. The coughing would wake them too, particularly when it rained. A chorus of retching and hacking identified each of the consumptives. The children would dread when it was their turn to collect and rinse the pans. They held their noses and couldn't look and had to be careful not to spill the fetid contents. As they got up so early, they could take themselves to the beach to throw off the stubborn smells and habits the guests brought into their house. If the tide was out, they would run so fast they thought even the wind wouldn't catch them. The middle two, Elizabeth and Albert, always headed straight to the fishing boats. They were rapt by the sense of adventure and the sea. Albert would go out with the fishermen when he was old enough, on a regular route north sometimes as far as Scotland to sell the fish. One day he didn't come back, not for three months. He settled in Annan just over the border as soon as his mam let him. Elizabeth missed him but married a fisherman on her sixteenth birthday and made sure Annie got a steady supply from the weekly catch.

This was the beginning. Forerunner to the rollercoaster ride, the celebrated myth of heyday that hid the price they paid. There were glorious times despite the grind of daily life. There was so much on offer at the seaside for Annie and her family, so many things to hide the deficits and the scarcity that sometimes took over. "The overlords", as Annie liked to say, bemoaned the "palaces of inferiority" that had sprung up everywhere. They made her blood boil, thinking they were better than everyone else. The season was short lived. It was only for a few short months that the thousands of visitors needed to be accommodated and taken care of. Annie and her family were gracious and grateful hosts; each weary or poorly guest reminding them of what they had left behind. Every moment was occupied with a smile, given gladly in service with half an eye on the winter months when their income would disappear and they had to find other ways to fend for themselves.

The First Layers—Yorkshire

Morecambe was the opposite of what they had come from: tightly packed terraces of squashed properties that housed and caged them. Damp and dark, the cottages belonged to the mills; the family belonged to them too, to the men who consumed them through their possession of them. They were plagued by consumption and thoroughly exhausted by it. Annie felt trapped by the closed stifling spaces she spent her life in. A short run from home to factory, already part of her, part of her family, the first layer of their map. Unfolding, it became their journey, about how they fitted and stayed, squeezed into buildings. Except, coughing and spluttering, they leaked at the edges and seeped through the cracks, always crawling towards the light. When they finally stepped outside their cramped cages, they soared. Only so far, but far enough to stop the decay: a full Wakes week in Morecambe, the holiday of a lifetime. "Beauty surrounds, health abounds": that's what the resort promised and true to its word, they were met by the salty breath of the Irish sea that filled their lungs while they scrubbed themselves half clean with coarse grains of sand. They had fought hard for the break. Annie thought of it as compensation, she felt it amounted to a lengthening of their time on earth, something the mill towns had tried their damnedest to cut short with their dust and fumes, the never-ending demands for labour that slowly bent and hollowed out their bodies. They wanted what they could never have: time. It didn't seem like a lot to ask: time to stop and look at themselves, to attempt to prise themselves away from early death, to breathe; a deep inhalation, one that quells the panic that sits at the well of us when pushed to the limit.

The Yorkshire mill workers were a full generation behind the cotton workers of Lancashire, the radical men and women who had organised and protested and showed them the way to the yearly exodus. Wakes week: a week's unpaid leave so the Overlords could keep their precocious machines running, so they could clean and repair their monsters. An unequal exchange. It was a sop of the highest order but it was all folk could hope for and it was almost worth it. Once they stepped off the train into the light of the seaside, they didn't miss the sound that drowned the noise of everything human—the sound of pain and panic until they became quietly subordinate. Morecambe gave escape from the enclosures that had been built around them without their consent. The seaside gave them what they didn't know they needed because they couldn't lift their heads long enough to see.

At the end of Wakes week, but a short time spent at the foot of sea and sand, they would return home not quite clean, not quite fresh, but cleaner, less weary, strong enough to keep working after they had removed the first layer of dust and partially cleared and filled their lungs. They returned—they always returned—pulled by a woollen umbilical cord back to their masters' because they were tied by work. Sometimes they lost a thread, momentarily just let it go to see how it felt, but they clung to it afterwards; flying between space and enclosure. The map, the route for so many, was a tapestry of containment and escape, east to west, to the light of the seaside.

The Second Layer—Sand, "The Quick"

Driven to move, Annie and her wayward family found themselves suddenly on the sand; hemmed in by the sea on one side and brick on the other. They had arrived but not yet formed, an emergent force, held at a threshold. Through their movement west, Annie had challenged who they were, the role they'd been allotted. They started to draw lines in the sand to restore themselves; a new map. Stood at the borderline of the quick, liminal creatures caught between migration and settlement. Unrepresented, the sand was the only place they could reside. It was their tentative benefactor (Fig. 10.2).

The sand
God's portent and
Keeper of left things

Fig. 10.2 Incoming tide, Morecambe © Richard Davis

The Bay was its own life force and danger came from movement; from tidal bores that roared, quietly at first but the sound grew and could knock you over. The tide came in from all directions as fast as a galloping horse, a constant rhythm and pattern of daily shifts. Sand and sediment were amassed, pushed into place by the force of incoming tides and the chaos caused by emptying rivers, which created channels for the sea to swallow and pools of quicksand to form. A full five rivers poured into Morecambe Bay adding to the danger and fascination, the pull of the watery realm and the muddy sand flats.

Annie felt as if she was at the edge of civilisation. She was continually forced to consider the precarity of her position, her loss and now her exposure and separation. With no one to confide in or help carry her load, Annie felt herself disturbed and distained through isolation. Standing at the periphery on the sand when the tide was out, she felt it keenly; she felt the whole weight of her predicament pressing into her. Annie had turned inwards, dulled her senses in order to hide the deep felt sorrow that threatened to swallow her. She had not had the time to say goodbye to her George and now the idea filled her with dread; her grief loomed as something physical at the edge of her vision but disappeared when she was on the sand. Fortuitously, Morecambe was the perfect place to spill herself. So much came tumbling out of her on to the sands; the enormity of it all, the sense of betrayal and rejection, her aloneness.

She'd been abandoned, she felt disgraced and left so alone with herself. Rage built up inside like a tower. Annie could feel it growing and taking over. She had to do something, find a way to bury it. She wanted to get rid of the way she felt but it was too rooted, it wouldn't let go. Annie had learned what to do though. She'd learned that the sand was a place to get rid of things, to bury them deep just by talking through them, whispering them out, forcing them under. The sand wasn't far, a stone's throw away, waiting for her. Expecting her. That's what she thought.

She remembers the first time she ran, fire feeding her legs, and as she dropped, she thrust her hands and arms down through the grainy wet of muddy sand, full force because the tide was far out; a big shouting sigh and a heave like she was a bellows expelling a whole factory floor full of air. She became tight and skinny and hard. She didn't sink like she thought she would, like sometimes she wanted to. She lay there, flat, stuck against the mud and the sand and she burned. The fire was right down the middle of her and she had to keep her mouth shut to keep it in, to stop herself from exploding or breathing it like a dragon. Air was one thing, fire was another. She didn't trust herself, or the rage she felt ripping through her. The way she landed hurt but she knew that what really hurt was shame and humiliation, it burned and raged, getting in the way. Annie felt her body become complicit, directed by the landscape and she became embarrassed because she behaved like a scolded child, an unhinged woman possessed.

It was like there was a flood, a thick rising tide that she couldn't wade through, stopping her in her tracks, pulling her under into quicksand. Annie tried to whisper but the words wouldn't come. She couldn't swallow them either. They stuck but she could never have spoken them anywhere else. Eventually, Annie unloaded some on to the sand and it felt healing so she took everything there. She buried her words, how they made her feel. She also found things there, a kind of strength, from calling and listening, and sometimes digging to hear other things. She found she could let go. The sand held her, literally, it absorbed the leakages she couldn't contain. They poured into it, but on their way, mingled with salt and coarse grains, they burned her skin.

Annie asked the sand to clear her and the sea to take away her outpourings, the things that weighed her down and ate her body. The contamination lingered at the edge of her breath, it seemed never quite to leave but she started to ask the sand to take it, to bury it and she became stronger. She would go to the sand to dump her load and she would watch from her vantage point as others like her did the same; casting off their ailments and contagion, the things they could no longer carry. Each stayed separate and did not draw attention to themselves. Sometimes there was an exchange, a brief look, an understanding statement spoken through eyes and the way someone held themselves. But friendships were rarely formed beyond the business of a guest house. There was simply no time, and not the habit (Fig. 10.3).

It was the men that usually died. The women didn't remarry but the men did. If it was the other way round, the men always remarried. There were many like Annie and her family; they carried the traces of ownership and toil. And even though they had moved, the habits of their former lives, for a while at least, continued to crush and clamour after them. Sooner or later though, their bodies succumbed to the power of the Bay, to bathing under its open skies, to the rhythm of the tides, and they began to feel settled, less harried. The sand was used as a holding place to deposit things, the things, when not up to the task, they couldn't quite reckon with. Annie could go back for them because she knew they were hers. And it wasn't easy to detach from a life, to leave it all behind no matter what pain was held there. Ultimately, everything was revealed or kept hidden according to the whim of the tide and the regurgitation of its underbelly. The Bay seemed to work with Annie, but she could not be sure.

Fig. 10.3   Low tide, Morecambe © Richard Davis

The Final Layers

Morecambe had become their saving grace. The family roamed the beaches and swam in the sea, their lines connected, drawn and pulled fast so they wouldn't go under. Annie knew they were safe because as she watched, she had learned to read the sand and the mood of the Bay; the way a wind and the sea's breath caught her, how the tides rolled in and the way the sands lay. Her children quickly found themselves at home in Morecambe, where they found freedom and the vestiges of a childhood denied to so many others.

Eventually, people came to Annie. Old Mr Ed, one of the fishermen that had taken Bessie and Albert under his wing, sent the first. He'd saved the woman from drowning after she'd become disoriented walking out onto the sands as the tides circled in. He'd practically carried her all the way back until he met Bessie and Albert who ran to get their mam. Annie restored the woman like old Ed knew she would. She brought the pale body round and administered her concoctions. Annie took the woman to the sands again the next day where she taught her to unload the burdens that had so confused and disoriented her. After that, others came. They just turned up, encouraged by the woman whose words had elevated Annie to something of a soothsayer. They were all looking for a healer but, in fact, it was the sands that drew them in. They would come in response to it but they were never sure why or what for until they walked onto the sand, urged or accompanied by Annie. Sometimes they needed her to watch as they sent their secrets and their burdens into it.

This was a reckoning that Annie hadn't seen coming. The sand was vast, it shifted and breathed and the sheer volume of space crushed her; for as far as the eye could see, she was surrounded by it. The sand accepted her unconditionally, and she stood as herself, unjudged. However, it was only when Annie found a cloth map on the sands that she fully saw herself. She had seen the map from a distance, moulded to the sand and recognised it straight away. She kept it: the keeper of left things.

Note

1. Oral history recorded in Morecambe July 2017.

11

"Over Sands to the Lakes": Journeys Over Morecambe Bay Before and After the Age of Steam

Christopher Donaldson

There are two notable clearings just north of Lancaster on the West Coast Main Line. The first of these is formed by the River Lune as it flows westwards towards Sunderland Point and the Irish Sea. The second is the broad level expanse of Morecambe Bay: that "majestic plain", as William Wordsworth has it, "whence the sea has retired".[1] This plain is no prairie. Beyond the samphire and cord grass that grows at its margins, its surface is almost bare and stretches, at ebb tide, for miles over the hard-packed flats of grit, silt, and mud known locally as "the sands". Twice daily, when the flood tide returns, it is swallowed by the sea. Yet for Wordsworth, who knew this plain as the southern gateway to the Lake District, it was both a passageway and a place set apart. "The Stranger", he writes, "from the moment he sets foot on those Sands, seems to leave the turmoil and traffic of the world behind him".[2]

For W. G. Collingwood, writing of the train journey around Morecambe Bay some seventy years later, the sentiment aroused by the sight of the sands was much the same:

> In old times, when I was a schoolboy, what a moment that was! After months in town, after the rush through Lancashire smoke, and the gradually lifting sky-line, slowly rising foreground of hedges and hills, at last – "Behold! – beyond!" as Ruskin said of his first sight of the Alps. [...] At Hest Bank we cross our frontier.[3]

C. Donaldson (✉)
Department of History, Bowland College, Lancaster University, Bailrigg, UK

© The Author(s) 2020
J. Carruthers and N. Dakkak (eds.), *Sandscapes*,
https://doi.org/10.1007/978-3-030-44780-9_11

163

Collingwood's invocation of John Ruskin (his late friend and teacher) intensifies the retrospective viewpoint that frames this passage. "Behold!—beyond!" is, after all, a quotation from *Praeterita*: that late, unfinished work wherein Ruskin recalls key places, people, and moments in his life.[4] This connection is worth noting as it deepens our appreciation of how, through Ruskin, a Wordsworthian spirit of remembrance enters into Collingwood's recollections. Here, we have Collingwood, the old man, casting a fond eye back to the Lakeland holidays of his boyhood: the journey over sands—a spot of time— calling to mind the memory of long summer rambles on the fells.[5] But it is not solely on this account that I have chosen to bring Wordsworth's and Collingwood's words together. What interests me more especially is the way they portray the sands of Morecambe Bay as a threshold, a "frontier": both a place passed over and a passage to a place beyond.

In this chapter, I shall examine this conception of the sands of Morecambe Bay by briefly tracing its reiteration in the works of a number of Wordsworth's and Collingwood's contemporaries. For the most part, I shall focus on the writings of the novelist Ann Radcliffe and of the journalist and poet Edwin Waugh. Along the way though, further consideration shall also be given to J. M. W. Turner's *Lancaster Sands* and to an early specimen of Elizabeth Gaskell's fiction. Collectively, as I shall show, the accounts provided by these artists bear witness to a period of remarkable change in human encounters and engagements with Morecambe Bay. Specifically, they offer insights into how the extension of a railway around the bay during the 1850s altered the way people experienced, understood, and represented the passage over this "majestic plain".

* * *

Morecambe Bay, as hinted above, lies along the north-western edge of Lancashire. Formerly, the Bay formed a division within Lancashire itself, separating the Hundred of Lonsdale into two parts: North of the Sands and South of the Sands. Today, it serves as a boundary between Lancashire and the modern county of Cumbria, into which the north of Lonsdale was incorporated nearly fifty years ago. A fluid, intertidal border, the margin of the Bay is a place of constant changefulness. It is notoriously difficult to map, as it reshapes itself with the turning of every tide (Fig. 11.1). It is also, paradoxically, both a centre and a periphery. Along its western edge, where it curls around what was once the secluded domain of the monks of Furness Abbey, the Bay reaches some of the remoter outposts of England. Yet, somewhere in the midst of this waterway lies the centroid—the geometric centre point—of the UK.

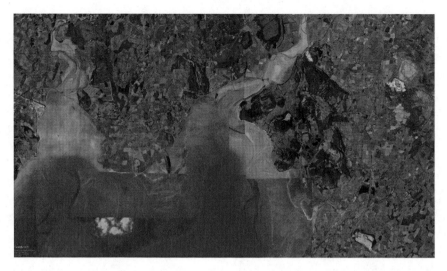

Fig. 11.1 Google Earth visualisation of the Kent Estuary in north Morecambe Bay (Note the incongruity between the tiled satellite photographs, each of which was captured on a different date. This disparity has been caused by the tidal shifting of the sands. *Source* 54°10′21.56″N and 2°51′46.33″W. *Google Earth.* 1 January 2004; 30 May 2009; 17 June 2010. Web. 1 May 2017)

The name *Morecambe* itself means "crooked sea", and this certainly goes some way towards describing the fallen arc of the Bay's foreshore, which scrolls northwards from Fleetwood to Heysham Harbour and around, past Humphrey Head, towards the Isle of Walney. This funnelled shape is formed by the confluence of the rivers Lune, Kent, and Leven, whose estuaries merge here with the waters of the rivers Keer, Wyre, and Winster to form one vast littoral plain. At flood tide, when the plain is submerged, it becomes an arm of the Irish Sea. At ebb tide, the waters recede to reveal a sub-country comprising some 120 square miles of inlets, mudflats, and intertidal channels.

The coastline around the Bay, for its part, mostly consists of marshlands, which have been shaped by shepherding and shore fishing for generations. On a clear day from the promenade at Morecambe (the Bay's old resort), you can survey the splendour of the shoreline in a single, sweeping glance. Gazing from the west, where the blocky outlines of Heysham Power Station rise on the horizon, you behold scattered towns and villages and, between them, rolling fields and pasturelands bordered by dry-stone walls. In the distance, a line of limestone ridges rises from the coastal plain, reaching heights of more than 700 feet at Hampsfell, above Grange-over-Sands, and Whitbarrow Scar, just beyond the A590. Towards the east, these steep outcrops give way to the rounded, semi-forested slopes of Arnside and Silverdale. Farther north, the grooved flanks of the Lakeland fells crest over the horizon.

In his guidebook *Over Sands to the Lakes*, Edwin Waugh portrayed this coastline as a panorama of "changeful picturesqueness". "Here", writes Waugh, "where the ragged selvedge of our mountain district softens into slopes of fertile beauty by the fitful sea,—and where the mountain streams, at last, wind silently homewards over the sands, we flit by many a sylvan nook, and many a country nest, where we should be glad to linger".[6] Waugh's words catch my attention because, although printed over 150 years ago, they are written from the perspective from which many travellers first see Morecambe Bay today: from the window of a railway carriage (Fig. 11.2).

Over Sands to the Lakes was published in 1860 (just three years after the completion of the Ulverstone & Lancaster Railway), and the book is in every sense a scenic guide to this new railway line. The book's subject is, as one might therefore expect, not so much Morecambe Bay itself, but the experience of journeying along the Bay by train. The view Waugh offers us is one framed by the window of a moving railway carriage. Consequently, in its finer flourishes, the style of his writing becomes almost cinematic. Although he occasionally pauses to sketch a specific scene in detail, his account is really a catalogue of passing glimpses and, in this way, conveys something of the curious power of railway travel to transform a landscape into a moving picture show. The passage I have just quoted—"we flit by many a sylvan nook"—is indicative, as is Waugh's description of the path of the line near Grange-over-Sands. Here, he writes: "The scenery richens [...] the grand bay on one side, on the other picturesque rocks and snatches of woodland, sloping to the shore, with the wild fells behind, all going by in panoramic flight".[7]

Even more intriguing than this, however, are those points at which Waugh pauses to inform his reader that the mobile spectacle unfolding outside the carriage is the result of not only the motion of the train, but also movement in the landscape. Thus his description of traversing the River Winster:

> The line now clips the rocky shore for about a mile, and we are rolling over the little river Winster, one of the boundary lines of Lancashire and Westmoreland [sic]; but, like the rest of these waters in Morecambe Bay, so changeful in its course over the sands, that yon pretty island, a little way from the shore [...] has been known to be first in Lancashire, then in Westmoreland, and back again in Lancashire, all in a month's time, through the caprice of this little Winster, which, when the fit is on it, thus plays at hide-and-seek with the two counties.[8]

At moments such as this, one comes to appreciate how Waugh's narrative figures the ceaseless movement of Morecambe Bay as integral to the traveller's enjoyment of the journey. The collocation of "little", "pretty", "play", and

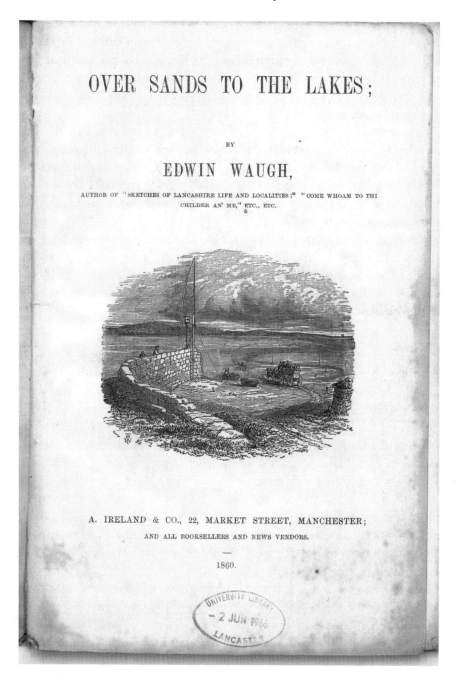

Fig. 11.2 Title page, Edwin Waugh, *Over Sands to the Lakes* (1860)

"caprice" in this passage—especially when combined with "hide-and-seek"—conspires to create a sense of leisurely pleasure-taking in the mobility of the terrain, and this pleasure is compounded by Waugh's mobile point of view. The train, as it rolls over the county boundary, increases the interest he takes in the correspondent flow of the Bay's waters—the one motion increasing the delight of the other.

What Waugh's account captures in such passages is a sense of the Bay's dynamic landscape character: its identity as a zone wherein the interplay of space and time can become dramatically apparent. Certainly, an appreciation of the dramatic spectacle of the Bay's shifting sands and tidal flows figures centrally in Waugh's book, but they do so largely to demonstrate his connoisseurship as a guide. In his hands, the mobile sandscape of the Bay becomes a curiosity at which to marvel from the enclosure of a railway carriage. But to focus exclusively on this aspect of Waugh's writing is to overlook another important feature of his book: namely, that as much as it celebrates the safe view of the sands afforded by the railway, it is also a swansong for an earlier and more adventurous age of trans-Bay travel.

* * *

Before the coming of the railway, as Waugh knew, the most direct way to travel to the Lakes from southern England was to follow the "over-sands" route: a journey of some twenty miles over land, sand, and sea, which involved fording both the Kent and Leven estuaries at low tide. Traversing this intertidal landscape could be treacherous, not least because of the Bay's quicksands and tidal bores. In poor weather, especially on misty evenings, the unwary and unfortunate have perished simply because they lost sight of the shore. But for many of the travellers of the pre-railway age, the potential danger of the over-sands crossing was part of its attraction. Indeed, as Waugh surmises, whether the trip was made on foot, on horseback, or by cart, the unpredictability of the sands gave "something of the piquancy of adventure" to the journey.[9]

True to Waugh's words, it is just this "piquancy of adventure" that comes across in many of the more famous accounts of the sands crossing from the pre-railway age. The most influential of these accounts was written by the novelist Ann Radcliffe. Today, Radcliffe is principally known as the "queen" of the golden age of the gothic novel. In her own time though, she was also admired for the sensory richness of her landscape writing. Her description of crossing Morecambe Bay, which appears in her *Journey Made in the Summer of 1794*, is exemplary. For Radcliffe, who travelled over Morecambe Bay by cart not long after sunrise, the passage over "these vast and desolate plains" was a singular experience made exhilarating by the slowly rising sea mists:

> The body of the sea, on the right, was still involved [in fog], and the distant mountains on our left, that crown the bay, were also viewless; but it was sublimely interesting to watch the heavy vapours beginning to move, then rolling in lengthening volumes over the scene, and, as they gradually dissipated, discovering through their veil the various objects they had concealed—fishermen with carts and nets stealing along the margin of the tide, little boats putting off from the shore, and, the view still enlarging as the vapours expanded, the main sea itself softening into the horizon, with here and there a dim sail moving in the hazy distance.[10]

The description is particularly noteworthy for the fitness of Radcliffe's style to her subject. In following the unfurling of this single, long sentence, the eye is made to enact, analogically, the sweeping glance that Radcliffe describes. This slowly expanding prospect of the sands contrasts the rapidly unfolding spectacle presented by Waugh. But even more striking than this is Radcliffe's attentiveness to the influence of the weather. Here, it is less the shifting nature of the terrain that shapes our sense of the scene than the movement of the "heavy vapours" that envelope the spectator. The suggestion of heaviness implies not only thickness, but also weight: the physical press of the atmosphere on the skin.

Drawing such distinctions helps to underscore a fundamental difference between Radcliffe's and Waugh's accounts, and it does not take much to see that this difference relates to the different modes of transport by which Radcliffe and Waugh were conveyed. Whereas for Waugh, who travelled around the Bay by train, the experience of the sands was of a landscape viewed from a secure enclosure, for Radcliffe, who crossed over the sands by cart, the experience was one of exposure. Indicatively, whereas the former, as a fast-moving observer, principally directs our attention to his movement through the landscape ("we flit by many a sylvan nook"), the latter mainly directs our attention to movement in the landscape itself. Notice how each of the "various objects" Radcliffe describes is distinguished by their motions: the "dim sails" and the mists *move*, and the latter also *roll*, *expand*, and *dissipate*; the fishermen and their carts *steal* along the shore; the sea *softens*.

* * *

The sensory richness of Radcliffe's account, and the sense of exposure it conveys, is characteristic of the records one finds of travel over Morecambe

Bay prior to the arrival of the railway. Individual writers varied in their appreciation of the experience. Some found it to be forbidding, others favourable or even fun. Much depended on the time of day and the weather. The Welsh naturalist and explorer Thomas Pennant, who crossed the bay on a "tempestuous evening", described the journey as "a melancholy ride of eleven miles" over "dreary wet sands, rendered more horrible by the approach of night" and "the driving of black clouds".[11] By contrast, William Wilberforce crossed the Bay on a bright September morning and found it to be a "very pleasant" and "most delightful" ride.[12] But whereas historical accounts of the Bay often differ in such particulars, they all affirm the assertions of William Cockin, who noted that the sense of immersion and isolation felt by travellers while crossing this "immense", "uniform", and "barren" landscape of water and sand could "affect [the mind] in a very sublime and unusual manner".[13]

The poet and topographical writer Norman Nicholson has written evocatively about the source of this "sublime and unusual" effect. "Out in the bay", he explains:

> The shore, with the line of cliffs, seems no taller than a bed of shingle, and even the fells themselves are dwarfed by the enormous flatness of the sands [...]. Here and there, skears and boulders and inexplicable old posts jut out of the sand, black with weeds and fortified with a whole Maginot Line of mussels and barnacles. There is an empty hollowness about the air, so that the clangings and pipings of the seabirds seem to echo one off another, having nothing else to echo off.[14]

This description of "flatness", "hollowness", and the echoing cries of the seabirds gives the sense of a desolate wilderness, and one made uncanny by the unexpected encounter with the detritus of civilisation: "old posts jut[ting] out of the sand". The picture Nicholson's words paint is not only one of isolation and the "inexplicable", but also one of potentially disorientating immensity, where one's perception of proximity, distance, and perspective is fundamentally altered.

Such a landscape, to borrow an expression from Robert Macfarlane's *The Old Ways*, might best be called unbiddable. Macfarlane uses the word in his account of the Broomway on the Essex marshes, and it seems equally applicable to Morecambe Bay.[15] Moving through such a place can cause one to look at the world rather differently. As Macfarlane puts it, such places "reconfigure local geographies, leaving known places outlandish or quickened". In this way, one might begin to perceive the relation between Cockin's account of the "sublime and unusual" effect of the sands and what the writer William L. Fox has called the experience of "cognitive dissonance in

Lancaster Sands

Fig. 11.3 Turner's *Lancaster Sands*, engraved by R. Brandard (J. M. W. Turner, *Picturesque Views in England and Wales. From Drawing by J. M. W. Turner* [London: Longman & Co., 1838], 19. *Source* Courtesy of the Cadbury Research Library, University of Birmingham, UK)

isotopic environments": the uncanny sensation that overcomes the mind in places whose "features are uniformly distributed in all directions".[16] In the case of landscapes such as Morecambe Bay, this notion of the uncanny—of the *unheimlich*, or unhomely—is, of course, particularly apt. Although human beings can occupy the sands for a short span of time, this landscape is ultimately inhospitable to our presence.

This sense of the fundamental inhospitality of the Bay is apparent in most historical accounts of the over-sands crossing, and it finds a noteworthy complement in the dozens of paintings and sketches of the Bay made prior to the arrival of the railway by artists including J. M. W. Turner, David Cox, and James Baker Pyne. Among these works, Turner's *Lancaster Sands* (1826), which depicts a coach, a cart, and two small parties of pedestrians moving shoreward, is particularly striking (Fig. 11.3).

When the painting is viewed as an engraving (the medium through which most of Turner's contemporaries would have known it), one can see all the more clearly how it pulls the eye along a series of diagonal lines. Some of these lines, such as the one made by the sun's reflection, represent natural phenomena. Others, such as the line of branches (or "brobs") marking the sand road, represent the works of humankind. Most of the lines apparent here, however, are a product of perspective. Among these we have not only the line from the tip of the coach driver's whip to the topmast of the distant ship, but also the line extending from the press of figures in the foreground to the cart in the distance which is charging shoreward. The effect of this composition pulls the eye successively from right to left and from foreground to background, creating a sense of both progression and recession.

The forward momentum of the figures, many of whom are doubled over with exertion, suggests strenuous, steady movement towards the shore, but without immediate evidence of danger. Notably, there are no breakers here to scatter the reflection of the setting sun. It is only when we begin to look closely that we see why the coach driver in the foreground—and, even more so, the cart driver in the middle distance—has raised his whip so earnestly. It is then that we begin to perceive that the receding line of brobs not only points to the path travelled by the advancing figures, but also signals the steady advance of the sea. This line, as it vanishes beneath the waters, indicates the disappearing path of safety and the approach of danger from beyond. As Ruskin observes in his reading of this engraving, the incidental detail in the lower left foreground of the dogs confronting the shrieking gull that advances admonishingly towards them suggests a moral for the entire scene. The "unexpected boldness" of this bird, writes Ruskin, "is a type of the anger of its ocean element, and warns us of the sea's advance".[17]

Ruskin's words are worth recalling as they emphasise how Turner presents Morecambe Bay as a place of perilous beauty. True to Turner's composition, the passage over the sands could be a dangerous undertaking. The deaths of the twenty-three Chinese cockle pickers who drowned after being abandoned near Hest Bank in February 2004 are a sombre reminder of a long list of fatalities extending back beyond the sixteenth century, the era from which the earliest accounts of the over-sands guides (or "Carters") have survived. The entries one finds in local records, such as the Registers of Cartmel Priory, present a grim roll call of the drowned:

1576. Sept. 12. One yong man wch was drowned in the brodwaters buryed
1577. Aug. 24. One little mann Rownd faced wch was Drouned at Grainge buryed
1582. Aug. 1. A son of Leonard Rallinsons of Furneis Fell drouned at The Grainge buryed
1597. Sept. 9. Henry Jnglish being drouned on Furnes Sand buryed

1597. Sept. 22. Richard Beeslaye beinge drouned vpon Kent Sands buryed.[18]

Yet, in spite of the potential hazard, before the coming of the railway in the 1850s, the over-sands route remained the main thoroughfare for travellers who wished to pass swiftly between southern England and the southern reaches of the English Lakes. The socio-economic consequences of the route were considerable, especially for towns such as Lancaster, which flourished as a regional transport hub between the late eighteenth century and the 1830s.[19]

Indeed, by the dawning of the age of picturesque tourism in the Lake District during the late eighteenth century, there were regular coach and carrier services running over the Bay throughout the week. An advertisement for one of these services published in the *Cumberland Pacquet* in September 1781 promised, for a fare of five shillings: a "sober and careful driver", and as "expeditious" a journey "as the tide will permit" from the Sun Inn, in Lancaster, to the King's Arms, in Ulverston.[20] Another announcement, included in Thomas Hartwell Horne's *The Lakes of Lancashire, Westmoreland, and Cumberland*, made a similar pledge, noting that whereas those "who shrink at the idea of the bay-crossing" could proceed to Ulverston via the Kendal and Ireleth turnpike, the over-sands route was half as long, half as expensive, and "certainly the most convenient for those who are in any degree restricted by time".[21]

Wordsworth, in the tenth book of *The Prelude*, provides a memorable description of the busyness of the sands during this period. Recollecting his journey over the Leven estuary in 1794, he describes the way before him as teaming with travellers. "All the plain", he writes:

> Lay spotted with a variegated crowd
> Of vehicles and travellers, horse and foot,
> Wading, beneath the conduct of their Guide,
> In loose procession through the shallow stream
> Of inland waters; the great sea meanwhile
> Heaved at safe distance, far retired. I paused,
> Longing for skill to paint a scene so bright
> And cheerful.[22]

Like the advertisements quoted above, these lines remind us of the importance of the sands as a place of commerce and traffic. But the emphasis Wordsworth places on the pleasantness of the "scene" before him reminds us of how, in spite of the association of the sands with danger, they were— then as now—also a place of visual delight. Little wonder, then, that he later commended the sands crossing to his fellow poet Felicia Hemans "not only [...] as a deed of 'derring do', but as a decided proof of taste".[23]

There were still fatal accidents, of course, many of which gave rise to local tales. Of these, the drowning of nine young men and women while returning home from a Whitsuntide Fair in 1846 is especially notable as it seems to have inspired one of Elizabeth Gaskell's earliest short stories, "The Sexton's Hero" (1847). Set in Silverdale, where Gaskell and her family holidayed regularly from 1843, the story blends the author's intimate knowledge of the scenery of the sands with a compelling portrait of local life and manners. The mise en scène with which the story begins is particularly striking for the way it invokes "the blue dazzle of Morecambe Bay" as a "sparkling" barrier dividing the churchyard, where the story's narrator and his friend are sitting, from "the more distant view".[24]

As the narrative unfolds, we realise that this view surveys the theatre of action for the inset narrative related to the narrator and his friend by an elderly Sexton, who joins their conversation. The Sexton's tale is a simple one. It concerns a young man named Gilbert Dawson, whom he had known in his youth when he lived in Lindale, a village on the far side of the Bay. Though the Sexton's eyesight has grown dim, he makes a point of directing the two friends to look towards the village on the distant shore:

> You can see Lindal [sic], sir, at evenings and mornings across the bay; a little to the right of Grange; at least, I used to see it, many a time and oft, afore my sight grew so dark: and I have spent many a quarter of an hour a-gazing at it far away, and thinking of the days I lived there, till the tears came so thick to my eyes, I could gaze no longer. I shall never look upon it again, either far off or near, but you may see it, both ways, and a terrible bonny spot it is.[25]

This passage is worth noting as it anticipates an important aspect of Gaskell's story, wherein the Sexton's failing eyesight is contrasted with his spiritual insight. Of even greater importance, though, is the subtle way that this passage conflates space and time. The distance of Lindal "across the bay" is merged with the temporal remoteness of the incidents the village's name recalls. The melancholy caused by the Sexton's memories is compounded by his inability to reach, or even to see, the farther shore. His body has grown too weak with age, his eyes too weary with regret.

As the narrative proceeds, it becomes evident that the Sexton is sharing a story from his early life, when as a young man he had competed with Dawson for the hand of a young woman, with whom he nearly drowned while returning over the sands from a wedding on the farther shore. The climax of his tale, which describes the moments just before the waters of the Bay overtake his cart, is worth quoting at length:

> By this time the mare was all in a lather, and trembling and panting, as if in mortal fright; for though we were on the last bank afore the second channel, the water was gathering up her legs; and she so tired out! When we came close to the channel she stood still, and not all my flogging could get her to stir; she fairly groaned aloud, and shook in a terrible quaking way.[...] I pulled out my knife to spur on the old mare, that it might end one way or the other, for by now the water was stealing sullenly up to the very axle-tree, let alone the white waves that knew no mercy in their steady advance. That one quarter of an hour, sir, seemed as long as all my life since. Thoughts, and fancies, and dreams, and memory ran into each other. The mist, the heavy mist, that was like a ghastly curtain, shutting us in for death, seemed to bring with it the scents of flowers that grew around our own threshold; it might be, for it was falling on them like blessed dew, though to us it was a shroud.[26]

What is remarkable about this passage is the tension Gaskell creates by contrasting the frenzy and panic of the narrator, and the terrified agony of the horse, with the almost imperceptible motion of the water "stealing sullenly" upon the cart. The violence and desperation of the Sexton's futile efforts to drive the mare forwards with his whip and then his knife are made all the more harrowing not merely by the "steady advance" of the bore tide (the "white waves that knew no mercy"), but also by the silent rise of the waters and descent of the mists. As readers we watch with him as every trace of the land, beneath and beyond, disappears. That, as his life flashes before his eyes, the last sense that remains to him is the scent of flowers at his own threshold underscores the funereal implications of the "ghastly curtain" the fog has wound round him. In this way, the whole of the passage works together to sharpen the suspense, until at the last minute the Sexton and his young wife are rescued by Gilbert Dawson—now revealed to be the story's eponymous hero—who gives his life in order to save them and whose corpse is later cast up on shore. The drowned mare and the ruined cart, we are told in conclusion, "were later found half-buried in a heap of sand".[27]

One comes across other local stories such as Gaskell's, including a fair number of tall tales that, if less religiose, are all the more entertaining for it. The itinerant lecturer and writer Adam Walker relates one in which a man uses his drowned horse as a buoy until he is able to light upon a sandbank tall enough to keep his head above the waves. "He stood up to his chin", writes Walker, "the waters went over him, he disengaged himself from his good friend the dead Horse, and waited there till the tide forsook the Sands, and got safe home".[28] The reader is left to wonder how the traveller managed to hold his footing against the rush of the tide. Waugh, for his part, published a story in which he and a friend narrowly escape drowning on the Leven sands. The tale concludes over dinner, which prompts the narrator to remark "I could not help feeling thankful that we were eating fish instead of being eaten by them".[29] Another yarn features an even finer bit of grim humour. A cautious visitor asks his guide if anyone has ever been lost on the sands, to which the guide chipperly responds: "I never knew of any lost [...]; there's one or two drowned now and then, but they're generally found somewhere i'th bed when th' tide goes out".[30]

The threat of death by drowning, a concern that ties these various tales together, is a theme that stands at variance with our typical notions of writing about the Lake District. There are exceptions one could cite here, of course: notably, the "drowned man" episode that Wordsworth included in the fifth book of *The Prelude*. More obscurely, one might look to Eliza Lynn Linton's now little-known novel *Lizzie Lorton of Greyrigg* (1866), whose eponymous heroine drowns herself in the fictional lake of Langthwaite. In the main, however, the risk of death by water is a danger on which Victorian accounts of the Lake District seldom touch. If the subject is raised, it is in the form of a story, which whether it be sombre or humorous is framed as a recollection of events that took place long ago. Even in Gaskell's "Sexton's Hero", the events of the Sexton's story are pushed into the distant past. They are supposed to have taken place some fifty years before.

What I want to propose in conclusion, then, is that this tendency in Victorian writing about the Lake District is, in part, a reflection of the new perspective of the Bay facilitated by the railway. It is true that the later nineteenth century would usher in a new mode of trans-Bay transportation in the form of paddle steamers, which became a popular seaside attraction in their time. But such innovations really only served to reinforce the shift in perception that the railway had introduced. Prior to the steam age, the Bay was a place over which to pass; the completion of the Ulverstone & Lancaster Railway made it a place all too easily passed over on the way to one's destination. The difference between the accounts of the sands in Radcliffe's *Journey* and Waugh's *Over Sands to the Lakes* suggests just as much. Whereas the former emphasises the exhilaration and exposure felt while travelling over the sands, the latter presents the Bay as a sort of scenic backdrop. This observation returns us to Collingwood's invocation of Ruskin's exclamation, "Behold!—beyond!". By the time Collingwood published that passage in 1902, the sand roads of Morecambe Bay had become a distant memory.

Suitably, in his transport, his attention is fixed not on the sands before him, but on the mountains in the distance—beyond.

Notes

1. William Wordsworth, *A Guide through the District of the Lakes in the North of England*, 5th ed. (Kendal: Hudson and Nicholson, 1835), ix.
2. Wordsworth, *A Guide*, ix.
3. W. G. Collingwood, *The Lake Counties* (London: J. W. Dent & Co., 1902), 1–2.
4. John Ruskin, *Praeterita and Dilecta*, ed. E. T. Cook and Alexander Wedderburn (London: George Allen, 1908), 115. Library Edition of the Works of John Ruskin, Vol. 35.
5. The phrase "spots of time" is Wordsworth's. See William Wordsworth, *The Prelude, or Growth of a Poet's Mind: An Autobiographical Poem* (London: Edward Moxon & Co., 1850), 325.
6. Edwin Waugh, *Over Sands to the Lakes* (Manchester: A. Ireland & Co., 1860), 6.
7. Waugh, *Over Sands to the Lakes*, 11.
8. Waugh, *Over Sands to the Lakes*, 11.
9. Waugh, *Over Sands to the Lakes*, 4.
10. Ann Radcliffe, *A Journey Made in the Summer of 1794*, 2 vols. (London: G. G. & J. Robinson, 1796), II, 497.
11. Thomas Pennant, *A Tour in Scotland and Voyage to the Hebrides MDCCLXXII*, 2 vols., 2nd ed. (Chester: J. Monk, 1774), I, 25–26.
12. William Wilberforce, *Journey to the Lake District from Cambridge 1779*, ed. C. E. Wrangham (London: Oriel Press, 1983), 46.
13. Thomas West, *A Guide to the Lakes, in Cumberland, Westmorland, and Lancashire*, 2nd ed., rev. by William Cockin (London: Richardson and Urquhart & J. Robson, 1780), 28.
14. Norman Nicholson, *Greater Lakeland* (London: R. Hale, 1969), 110.
15. Robert Macfarlane, *The Old Ways: A Journey on Foot* (London: Penguin Hamish Hamilton, 2012), 78.
16. Macfarlane, *The Old Ways*, 78; William L. Fox, *The Black Rock Desert* (Tucson: University of Arizona Press, 2009), 20; see also Fox's "Walking in Circles: Cognition and Science in High Places", in *High Places*, ed. D. Cosgrove and V. della Dora (London: I.B. Tauris & Co. Ltd., 2002), 19–32.
17. John Ruskin, *The Elements of Drawing; in Three Letters to Beginners* (London: Smith, Elder, & Co., 1857), 325.
18. Henry Brierley, Amy Wilson, and Clement Hall Brierley, *The Registers of the Parish Church of Cartmel in the County of Lancaster* (Rochdale: James Clegg, 1907), 128ff.

19. See James Bowen, "The carriers of Lancaster 1824–1912", *The Local Historian* no. 40 (2010): 178–190.

20. John Fell, "Guides over the Kent and Leven Sands, Morecambe Bay", *Transactions of the Cumberland and Westmorland Antiquarian and Archaeological Society*, no. 7 (1884): 5–6.

21. Thomas Hartwell Horne, *The Lakes of Lancashire, Westmoreland, and Cumberland; Delineated in Forty-Three Engravings from Drawings by Joseph Farington, R.A.* (London: T. Cadell & W. Davies, 1816), 4.

22. Wordsworth, *The Prelude*, 290.

23. Harriet Brown (ed.), *The Works of Felicia Hemans, With a Memoir of Her Life*, 7 vols. (Edinburgh: William Blackwood & Sons; London: Thomas Cadell, 1839), I, 207–208.

24. Elizabeth Gaskell, "The Sexton's Hero", *Howitt's Journal*, no. 2 (1847): 149.

25. Gaskell, "The Sexton's Hero", 150.

26. Gaskell, "The Sexton's Hero", 151.

27. Gaskell, "The Sexton's Hero", 152.

28. Adam Walker, *Remarks Made in a Tour from London to the Lakes of Westmoreland and Cumberland in the Summer of MDCCXCI* (London: G. Nicol, 1792), 55.

29. Edwin Waugh, "Chapel Island; Or, an Adventure on the Ulverstone Sands", *Lancashire Sketches*, 3rd ed. (London: Simpkin, Marshall & Co., 1869), 15.

30. Waugh, *Over Sands to the Lakes*, 4.

12

Sands Immense: A Fool's Errand

Jean Sprackland

Day One: Southport

The path is steep and the loose sand creaks and slips under the soles of my boots. Something skitters away in front of me—rabbit, probably—and as it dashes to safety, I see the grey-green heads of the marram grass flick aside and then back into place, like hair parted briefly in a breeze. Getting off the beach into the dunes is like escaping from a busy street into a maze of quiet lanes, where time seems to run more slowly. The traffic-roar of the sea, the hectic wind and the screaming of the gulls are still audible here, but softened and muted. Other sounds come into focus: the pebble-tap of a wheatear, the rhythm of my own breath, the sigh the dry sand makes as it collapses and fills my footprints. I leave little trace as I pass.

I'm following in the long-vanished footsteps of two men who took this path, or another like it, a hundred and sixty years ahead of me. One was wild-eyed and thickly bearded, unusually sunburnt for a winter day; the other was older, more wary-looking, sporting a fashionable moustache and wearing a frock coat, warm but rather formal for the beach. The first was Herman Melville and the second Nathaniel Hawthorne, and they were out for a walk together along the shore from Southport. It was a November afternoon, and there was a cold wind blowing in off the sea. Darkness would come on at about four o'clock, after a scant eight hours of daylight. They strode out, despite the weather and the shortness of the day; both were strong walkers,

J. Sprackland (✉)
Manchester Metropolitan University, Manchester, UK

© The Author(s) 2020
J. Carruthers and N. Dakkak (eds.), *Sandscapes*,
https://doi.org/10.1007/978-3-030-44780-9_12

and they covered six or seven miles easily. "An agreeable day", wrote Melville in his journal that night. "Took a long walk by the sea. Sands & grass. Wild & desolate. A strong wind. Good talk. In the evening Stout at Fox & Geese".[1]

What were they doing here, these two giants of American literature? They are unlikely figures here in this English watering-place, not yet fashionable, but beginning to acquire a few brisk attractions: donkey-cart rides, boat trips and draughty boarding-houses. It has the quality of myth, this double visitation. I lived here for twenty years, and in all that time I never heard anyone mention it, though small towns are generally proud of such associations.

It started with that good strong wind. Hawthorne had taken up the post of American Consul in Liverpool, a breezy enough place in its own right, but when his wife's health faltered the medical advice was that she needed to be out of town, in the bracing air that swept in off the sea at Southport, clean and invigorating. So the family rented a suite of rooms in a terraced house on the promenade, and Nathaniel commuted by train to Liverpool each day to fulfil the duties of his office. His first impressions were not favourable. "It is the strangest place to come for the pleasures of the sea", he complained, "nothing but sand-hillocks covered with coarse grass".[2]

Melville's visit was a brief one, an early stop on a long tour which would take him through Europe and the Middle East. His in-laws had provided the money to pay for the trip, out of concern for his wife, who was finding it difficult to cope with his mood swings and unpredictable behaviour. He was eager to visit Liverpool and renew his connection with Hawthorne. To call this connection a friendship would be like calling the sea a big puddle; Melville worshipped Hawthorne and idealised the bond between them. He once wrote him a letter in which he said "Your heart beat in my ribs and mine in yours, and both in God's…".[3] If Hawthorne welcomed this devotion to begin with, it grew increasingly uncomfortable as time went on. Melville was a turbulent and needy soul, and the relationship was out of balance.

I have come back to this town for the first time in several years, on a mission to retrace the two men's steps along the beach and into the dunes where they sat down and talked. I have three days in which to track them over the sand. During my years here, I spent a lot of time doing as they did that day, sharing the very same spaces without knowing it. I'm responding to that powerful human impulse: the desire to locate the site of an event, to be physically present in the same spot where it happened, even if no visible trace remains. To stand in a field of corn which was once a battlefield, or in a city square where a revolution began, or in a pub where a rock star once propped up the bar. I feel it now, the longing of the pilgrim.

The row of terraced houses on the promenade where the Hawthornes lived was demolished long ago to make way for a hotel, and now that in turn has been replaced by a car park. Some of the other mid-nineteenth-century houses still stand, however, with an air of bashed and weary grandeur. The prom remains a broad, sweeping highway, where people stroll, as Hawthorne once complained, "without any imaginable object".[4] But it has experienced a dislocation; it has lost its proximity to the seafront, and the houses here no longer command a view of the beach. By the early twentieth century, the sea had receded so far that the decision was taken to reclaim some of the foreshore and use it to lay out parks, gardens, a large lake, fairground and floral hall. The house where Hawthorne's daughter Una liked to sit on the window-seat watching the sea on rough days is now a fifteen-minute walk away from the beach: through King's Gardens, across the ornamental bridge over the lake, and past the model village with its ice cream stand and narrow-gauge railway. I wonder what Hawthorne would have made of it. Even in 1856 he was claiming with grumpy hyperbole: "In all my experience of Southport, I have not yet seen the sea".[5] He and Una can't both have been right. But he had a point: when I stood at the far end of the pier this morning, I gazed down and saw bare sand stretching away beneath.

From where I stood, beyond the pier cafe and the coin-slot telescope machine, near the place where fishermen gather when the tide is high enough, I looked back over the great flat expanse of the beach, south towards Smith's Slacks where the embryo dunes are rising. There the beach is green with vegetation, which has acted as a system of traps for blown sand. The sand accumulates into tiny peaks, which grow and are colonised by marram and other plants and gradually become recognisable as sand dunes. Out of view from here, a couple of miles further down the coast at Ainsdale, you can see where this process leads. It's one of the largest dune systems in Europe, a place where the same accumulation has been going on for centuries, resulting in a dramatic sandscape which provides a complex series of habitats for some of our most endangered plants and animals. That's where I like to imagine that Hawthorne and Melville ended up, on their "good long walk". In 1856, there wasn't much of a village there, just a scattering of small farms, rabbit warrens and fishing boats. The sand dunes were not treasured and protected but characterised as "barren" and "unproductive", and the main concern was to find ways to change and control them, so that they could be brought into cultivation. A few miles down the coast at Freshfield, an experiment was underway, using the new railway to bring human sewage from Liverpool to fertilise the soil there and raise potatoes and asparagus; close to the station there was even a special "manure siding" for the purpose. But at Ainsdale the dunes were wild and remote and uncultivated, and they feel that way still.

Today is one of those sharp, ecstatic days you get here—the light and the wind and the flung-open sky, salt on my lips, blood fizzing in my veins. The pleasure of being here, with only the seabirds for company, and that sense of life as something not owned but coursing through me as it does through them. I love this place, I love walking here. But it wasn't always so. When I came to live here twenty-five years ago, I felt washed-up, beached and hopelessly out of place. This was not my idea of the seaside at all. The shore was bare and bleak, the sea distant and flat and grey. The sandhills just seemed empty and featureless. Looking back on that time, I can forgive Hawthorne, who complained that Southport was "as stupid a place as ever I lived in".[6] No wonder the local tourist board doesn't try to make capital from the association. I was not much troubled, as he was, by the paucity of baronets in the town, or the "tradesmanlike air" of the promenade, but I do feel a flash of recognition when I read this entry from his journal: "I cannot but bewail my ill fortune, to have been compelled to spend these many months on these barren sands, when almost every other square yard of England contains something that would have been historically or poetically interesting. Our life here has been a blank".[7] I felt a bit like that myself once. He lived here only for one year, and it took me much longer than that to learn how to look at the place. Then, it was like seeing it through a different lens: what had seemed empty and barren was vividly and tumultuously alive, in ways I simply had not been able to understand.

I've scrambled up a steep dune and arrived at the summit. Look, Nathaniel! Can't you see how dynamic this place is? With the fervour of the convert, I want him to see it the way I do: stripped to the simplicity of three basic elements—water, air and sand—which interact all the time in a continual state of flux. I want him to notice that the beach is different every time you go there: its surface ridged or smooth, sometimes with high berms you can climb and walk along, sometimes sliced through with deep channels of water. And the restlessness of the dunescape, shifting, settling, reconfiguring. I want him to get it, to recognise that this is a landscape with character: wild, mercurial, essentially unknowable. These are things you ought to understand, Nathaniel Hawthorne, with your dark romanticism, your fascination with that indeterminate space which lies between the dream and the material. I want you to see what I learned to see, when my love affair with this place began.

Day Two: Intimacies

A thundery darkness is beginning to gather at the edge of the sky; there's going to be a downpour. I'm walking barefoot, sandals in hand. In contrast to the cool, damp undertow you can feel when you walk where the tide has been, the sand here in the dunes is almost too hot to walk on.

Whatever can Herman Melville—who as a young man joined a whaling expedition, sailed the high seas and jumped ship in the South Pacific—have made of this place? Walking where I walked today, with a broad vista of empty sand and a flat grey ribbon of sea on the horizon, he must surely have found it all very tame. At a squint, there might conceivably have been something to remind him of the island of Nantucket, which he has Ishmael describe as "a mere hillock and elbow of sand; all beach, without a background".[8] His brief journal entry gives no clue. He notes only that the day was "agreeable", the conversation "good".

His visit seems to have taken the Hawthornes by surprise, but he was welcomed generously if not enthusiastically and spent a few days with the family in Southport, causing some ripples of concern with his inadequate luggage and relaxed attitude to personal hygiene. The walk along the beach was the first chance for the two men to talk at length, and even then, the really significant stuff was kept back until they turned inland, found themselves a dip in the sand dunes and sat down to rest. There, at last, was the opportunity Melville had been hoping for, the moment in which he could unburden himself and share his clamorous thoughts with the man he thought of as his kindred spirit. "Sat down in a hollow among the sand hills (sheltering ourselves from the high, cool wind) and smoked a cigar", wrote Hawthorne in his journal. "Melville, as he always does, began to reason of Providence and futurity, and of everything that lies beyond human ken, and informed me that he had 'pretty much made up his mind to be annihilated'; but still he does not seem to rest in that anticipation; and, I think, will never rest until he gets hold of a definite belief".[9] It certainly wasn't the scenery Melville had come for—indeed he may scarcely have noticed it—but the chance to grapple with the questions and terrors that were obsessing him. The chance to say out loud, possibly for the first time, that all he could see ahead of him was oblivion.

As it happened, the hollow in the dunes provided just the right conditions for these painful confidences. Sand dunes offer shelter, protection and privacy. They adjoin the liminal, open space of the shore, but are fortified against it. A retreat here is an opportunity to escape or transcend the usual constraints of time and place. Sand is so mobile, and the dunescape so shifting and changing, that any peak or valley is inevitably temporary. I'm not talking about change over geological time, but a remaking of the landscape that can be observed over the course of days, if the weather is rough. This gives the dip where you sit an ephemeral quality—it feels less like a *place* in the usual sense and more like a den, a tent, or perhaps a boat at anchor. While you are here, you exist outside the current of human events. Meanwhile, all around you, events in the lives of everything else—natterjack toad, sand lizard, skylark—continue with the same urgency as ever. You feel a sense of continuity and an awareness of transience, and the two experienced together are the nearest I can get to a definition of *peace*.

When I lived here, I walked in these dunes three or four times a week, and on those walks, I would often experience that sense of safe haven. I was usually alone, but not always. I sat once with a friend, sheltering from the wind just as Hawthorne and Melville did, not smoking a cigar but sharing an apple and some cheese. I was slicing the apple with a rather blunt penknife, and my friend suddenly said: "It's like me hacking away at the prozac with a razor blade". She was trying to ease herself down to a lower dose, she said, but things kept getting bad again. So she'd started customising the pills, shaving a little bit more off each day, experimenting, watching her own response to see what she could get away with. If she shaved off too much, she told me, she would think of killing herself not just at night when she was trying to go to sleep, which was normal, but also while she was washing up or eating ice cream or watching TV with the kids. She could be laughing at *The Simpsons* and thinking of killing herself, both at the same time. I had no idea. I'd thought we were close, but she hadn't told me any of this. I suppose she might never have told me, if we hadn't been tucked away in this warm little booth in the dunes, walled off from earshot: a safe, intimate space like a confessional in which secrets could be shared.

I've used that same confessional many times myself, even when alone. During the darkest times of my life, walking on the beach became more essential than ever, because I could talk to myself there. "Oh what joy for a shy man to feel himself so solitary that he may lift his voice to the highest pitch without hazard of a listener!" wrote Hawthorne,[10] and out on the lonely, windswept shore I have been free to rage and weep and howl, overheard only by flocks of oystercatchers and sandpipers and the occasional dispassionate cormorant. Then to creep exhausted into the dunes and tuck myself away from the world, and to talk myself back into one piece, laying out the facts, going through the pros and cons, feeding myself little titbits of wisdom or comfort or resolve before I set off home to face the music again.

In these remote and private places, there are sometimes intimacies of another kind. On one of my solitary walks a few years ago, I stumbled on a naked couple in the dunes, in a hidden dip where the wind couldn't touch them. They quickly scuffled together, rather than springing apart, and she hid her face against his shoulder. But he stared at me, fierce and straight. It was as if the place was under an enchantment, and I had blundered in and was caught there for a moment. I seemed unable to look away. I remember he was propped on his elbows above her, and the hairs on his arms were thick with sand. I was ambushed by the sight of the two of them, sunlit and archetypal, in the creel of warmth that sheltered them as such places must have sheltered many lovers over the centuries. I watched a damselfly dart in and pause on her discarded underwear, laser blue on faded black. Silence. Then a skylark started up like a machine. "Well?" he said, in a voice like a snapped twig, and the spell was broken and I stepped back and strode quickly away.

Now the first few huge drops of rain test the sand silently, splashing it with sudden contrast, so that it's not one thing but many: aggregate of ancient rocks, sea-creatures, glitter of minerals, bright sift of plastic. The stuff of metaphor, reminding us, generation after generation, of the ephemeral nature of our lives, our insignificance in the context of geological time. The footprints we leave, or the name scratched with a finger and lost with the next tide. Here in the dunes, secrets are shared, lovers meet, and the unspeakable is spoken. The sand doesn't care, is on the move anyway. Whatever fragments or traces of those exchanges might remain, they'll be scattered and lost next time the wind blows. And now the rain is yanked down hard like a shutter. Smell of seaweed and petrichor, and the sand stained dark.

Day Three: Evidence

The wind pummels my back and rattles the hood of my jacket as I turn inland with relief. The sense of timelessness is tangible this afternoon; I half-expect to hear the two men's voices, and be able to track them down to where they are sitting in their heavy coats under the November sky, one talking on in melancholy loops and circles, the other smoking, shaking his head, uncomfortably aware of the cold in his bones and the afternoon light beginning to fade.

On the day of the walk, Herman Melville was thirty-five years old, and feeling finished. It was five years since the publication of *Moby Dick*; he had poured himself into that work and it had been a flop. In fact, his only real commercial and critical success was long behind him: his first novel, *Typee*, a tale based on the time he spent living on a Polynesian island. Now he was struggling, pulled in two opposite directions—feeling he was expected to replicate his youthful success with another crowd-pleaser, and wanting desperately to write the books that really mattered to him, knowing they were unlikely to make any money or enhance his literary reputation. "All fame is patronage", he wrote. "Let me be infamous; there is no patronage in that".[11] They are brave words, but really he was afraid of ending up known only as "a man who lived among the cannibals". Meanwhile, his inner life was tumultuous; he made himself and those around him miserable with moods and tempers and incessant questions of religious faith and doubt. "It is strange", wrote Hawthorne that evening, "how he persists – and has persisted ever since I knew him, and probably long before – in wandering to-and-fro over these deserts, as dismal and monotonous as the sand hills amid which we were sitting. He can neither believe, nor be comfortable in his unbelief; and he is too honest and courageous not to try to do one or the other".[12]

Deserts came to matter very much to Melville. After he left Southport and Hawthorne behind, he would travel on to the Holy Land, an experience which affected him deeply and gave rise to his 18,000-line verse narrative *Clarel*. The sandy landscapes of the desert, so important in the poem, are not seen as dismal or monotonous. At first, they are perceived to have "a charm, a beauty from the heaven / Above them"; they are reminiscent of "Western counties all in grain / Ripe for the sickleman and wain". But this bucolic image is superseded by something much more characteristic of Melville. The desert, observed over time, starts to take on some of the power and significance that in his previous work has pertained to the sea:

> Sands immense
> Impart the oceanic sense:
> The flying grit like scud is made:
> Pillars of sand which whirl about
> Or arc along in colonnade,
> True kin be to the waterspout.
> Yonder on the horizon, red
> With storm, see there the caravan
> Straggling long-drawn, dispirited;
> Mark how it labors like a fleet
> Dismasted, which the cross-winds fan
> In crippled disaster of retreat
> From battle.[13]

Of course he's talking about the Egyptian desert, not the Lancashire coast. But Herman, look at this place today—the movement is just as you describe it, the flying and whirling scud of sand, like waves and spray and spume on a stormy day at sea. Sand, with its perpetual movement, its changing aspect, its restlessness and inscrutability, becomes in this strange poem of yours an element as contradictory and as terrifying as the question of faith itself. Can Clarel and his fellow pilgrims imagine religious faith surviving and coexisting somehow with science in the age of Darwin? Or are they now confronted with the reality of a godless universe, where "unperturbed over deserts riven, / Stretched the clear vault of hollow heaven"?[14] Sand creates a landscape whose very emptiness is charged with meaning, just as the ocean is charged with meaning for Captain Ahab and his crew, even if ultimately what the meaning boils down to is *meaninglessness*.

The sand where they sat, Melville ruminating, Hawthorne fidgeting—where is it now? The wind has blown it, picked it up and flung it apart, over and over again, in the fifty-eight thousand days since. The dunes have grown and fallen, like houses, like empires. The sea has retreated, and the place has been covered over with flowerbeds or tarmac. Their words were taken into the air and scattered, of course, but so was the location in which they were spoken. Even if Melville had written the coordinates in his journal, I would not be able to find the place. It is *nowhere*. On this coast, topographical change is not just something you read about in books, or extrapolate from the landscape; you observe it happening with your own eyes, sometimes literally overnight. It makes for a different experience of environment; nothing is fixed, nothing is stable. In the past, this instability was a recurring threat to homes, livelihoods and everything people had. Storms could bring inundation by water, but also by sand. Every so often a church, a farmstead, even a whole settlement was buried in a catastrophic sandblow. The houses and streets and middens of Argarmeols or Ravenmeols, both engulfed long ago, are buried somewhere under the dunes, maybe right beneath the spot where the two men sat. For the inhabitants of those lost villages, landscape change was not a matter of idle interest but an immediate preoccupation, a permanent state of crisis which they tried to mitigate by using all sorts of methods of holding the sand firm. The need to introduce a different texture and structure, to stop it blowing about, provided arguments for various enterprises over the centuries, from rabbit farming to the dumping of spent tobacco waste. "Star" or marram grass, which grows readily in sand and has tough roots that are good at trapping it, was so important here that watchmen were appointed to supervise its planting and to watch out for anyone breaking the law by cutting it. The experimental planting of pine trees to stabilise the dunes began at the end of the eighteenth century, and the plantations are now so well established that they seem to have been here forever. Those pine trees are, quite literally, rooted in the past: when the saplings were planted, the holes were filled with a rich growing medium known as "sea slutch". It was black silt, dug from the foreshore at Formby, now known to contain Neolithic human and animal footprints. Even prehistory was enlisted in the battle to bring the dunes under control, a battle which may have some local success for a time but which is ultimately a lost cause.

Lost causes and fools' errands. No pilgrimage can be made to the place where Hawthorne and Melville sat and talked on that November day, because it no longer exists. When I say it's nowhere, I mean it's *everywhere*: atomised, spread like seed on the wind. I'm digging with my gloved hand as I sit and think about this, turning over the sand, as if in spite of everything I know I might find the evidence. Come on, let there be *something*. What the sand buries, it eventually disgorges, after all; what is concealed is in time revealed. This dynamic stretch of coast is now in a period of marine transgression. The sea is encroaching, breaking down the frontal dunes and exposing a buried landscape which has been sealed away for many years. Medieval trackways, Victorian shipwrecks, 1960s caravans—the past is being laid bare again, long after everyone had forgotten.

Did Melville, a man possessed by metaphor, see any of this? Did he recognise his own restlessness in this place, the concealment and rediscovery, the way his buried thoughts made their way again and again to the surface? A dirty glint, worked free, turns out to be a tiny pocket-knife, rust-pocked and with a faded green handle. I pick it up and rub the sand off it, wanting to kid myself it's the one Nathaniel Hawthorne used to peel an apple, just as I did when I sat here with my friend a few years back. But the handle is plastic, stamped with the legend CONROY LTD. Besides, a knife is way too ambitious. I'd be happy with a lot less. A few shreds of tobacco from Herman Melville's cigar, perhaps. A shell he picked up and admired. A grain of warm sand he held, with thousands of others, and allowed to run away between his fingers. I scoop again and examine my cupped palm.

Notes

1. *Journals*, eds. Howard C Horsford and Lynn Horth (Evanston: Northwestern University Press, 1989), 51.
2. Nathaniel Hawthorne, *Passages from the English Notebooks*, accessed 2 December 2019, http://public-library.uk/ebooks/32/42.pdf.
3. "Melville's Letters to Hawthorne", Letter 7, *The Life and Works of Herman Melville*, accessed 2 December 2019, http://www.melville.org/letter7.htm.
4. Hawthorne, *Passages from the English Notebooks*.

5. Hawthorne, *Passages from the English Notebooks*.
6. Hawthorne, *Passages from the English Notebooks*.
7. Hawthorne, *Passages from the English Notebooks*.
8. Herman Melville, *Moby Dick* (London: Vintage, 2007), 68.
9. Hawthorne, *Passages from the English Notebooks*.
10. Nathaniel Hawthorne, "Foot-Prints on the Sea Shore", in *Tales and Sketches* (New York: Library of America, 1982), 568.
11. "Melville's Letters to Hawthorne", Letter 3, *The Life and Works of Herman Melville*, accessed 2 December 2019, http://www.melville.org/letter3.htm.
12. Hawthorne, *Passages from the English Notebooks*.
13. Herman Melville, *Selected Poems* (New York: Penguin, 2006), 158.
14. Melville, *Selected Poems*, 180.

13

Confounding Cartography: The Sandscape Diminution of Hayling Island

David Cooper and Michelle Green

NOTICE TO ALL KITESURFERS
Private Beach
This is a private beach owned by Hayling Golf Club, and licenced to CBK (Hayling Island) Ltd for kitesurfing by its club members and customers only. If you are not a member of CBK you may not kitesurf at this beach. Please contact the Beachlands Office (opposite the funfair) for details of public kitesurfing sites on the island.[1]

* * *

Hayling is an island in the county of Hampshire, southern England. Separated from the mainland by a narrow channel, this flatscape is sandwiched between the tidal mudflats and salt marshes of Langstone and Chichester Harbours. It is an island thick with stories. Archaeological excavations in the north have revealed an Iron Age temple that was succeeded by a Romano-Celtic place of worship and, later, a Saxon ritual site. Moving further south, it is said that a grave in St Mary's Church marks the entrance to a smugglers' tunnel leading down to the shore. In 1932, Catherine Yurievskaya—the daughter of Alexander II of Russia—retreated to Hayling after a career as a professional singer; and, twelve years later, the island was the site for a

D. Cooper (✉)
Manchester Metropolitan University, Manchester, UK

M. Green
Manchester, UK

© The Author(s) 2020
J. Carruthers and N. Dakkak (eds.), *Sandscapes*,
https://doi.org/10.1007/978-3-030-44780-9_13

mock invasion exercise—Fabius 2—in preparation for the D-Day landings. The island has a rich spatial history. Yet it also has a distressingly precarious future. Low-lying Hayling—as with so many islands—is in a position of acute vulnerability in this age of climate emergency.

Hayling, therefore, could conceivably serve as a key reference point for the Anthropocenic imagination. Curiously, however, it doesn't feature on the contemporary literary map. The preoccupation with place in twenty-first-century literary culture has led to an interrogation of Britain's archipelagic geography. This obsession is evident in *Islander: A Journey Around Our Archipelago*: a popular creative non-fictional text in which the nature writer, Patrick Barkham, maps out eleven small islands around the British and Irish coast.[2] It is similarly evident in critical scholarship since, following the publication of John Kerrigan's *Archipelagic English: Literature, History, and Politics 1603–1707*, the framing of Britain and Ireland as an archipelago has informed literary geographical analyses of a heterogeneous range of place-specific texts.[3] Yet, Hayling has not featured in any of the major creative and/or critical writings that have appeared as part of this archipelagic turn. When it comes to the literary culture of the present, this island remains surprisingly uncharted territory.

There are several possible, imbricated reasons for this cultural marginalisation. First, the archipelagic project has been informed—explicitly and implicitly—by a twenty-first-century imperative to reconfigure the geopolitical relationships between England, Ireland, Scotland and Wales. In an article examining the "new nature writing", Jos Smith argues that most archipelagic reimaginings have focused on "a decidedly Celtic fringe and a perhaps more-than-partially devolved cultural geography".[4] In thinking archipelagically, the most popular and influential creative non-fiction writers—from Robert Macfarlane to Kathleen Jamie to Tim Robinson—have gravitated to the north and the west. To apply Kerrigan's literary critical thinking, then, Hayling's geographical location within the archipelago means that it is inextricably implicated within the paradigm of Anglocentrism—centred, of course, in the capital—that ought to be questioned and stripped away by twenty-first-century place writers and literary geographers.[5] In this analysis, Hayling—less than 100 miles away from Waterloo train station—is just too close to London.

Given that Barkham enthusiastically explores the Essex islands of Osea and Ray, however, the relative proximity to London can only be part of the story. Hayling's cultural ostracisation is perhaps also attributable to the accessibility of the island. As Stewart Williams points out: "there is a common conception (or rather misconception) with the pictorial representation of islands erring towards a perfect, totalizing circumscription of space".[6] In the popular imagination, an island is a landmass—almost invariably circular in nature—absolutely bounded by water. Hayling, though, falls short of this ideal as a bridge to the mainland was built in 1824.[7] Except on the spring tide, the modern visitor's passage onto Hayling is unremarkable as the car smoothly progresses along the A3023 that links the coastal village of Langstone with the northern end of the island. The bridge road carries you over Sweare Deep; but, even at thirty miles per hour, blink and you'll miss it.

Hayling's marginalisation may also be due to the material and social geographies of the island itself. Within its nineteen square miles, Hayling contains a range of landscapes divided into two parishes.[8] The thinly populated North Hayling is characterised by a jumble of arable farmland, barren wasteland and marshes. Both agricultural land and marshy wastes are similarly found in the southern half of the island. South Hayling, however, also holds suburban estates and parades of shops; marinas and beaches; a funfair and sites of special scientific interest. Hayling is a relatively small island in which a wide range of spaces and spatial histories coalesce. Ultimately, this complexity means that Hayling resists the kind of single, monolithic place-narrative that is associated with so many islands of the imagination.

* * *

Havant Borough Council, "Hayling Island Seafront Regeneration Analysis and Feasibility Study"
    [...] Eastoke beach is a steep shingle, groyned beach which features the start of a promenade that runs through to the nature reserve and sailing club at Sandy Point. Facilities include car parking, toilets, beach railway, outdoor gym and playground and easy access to shops.[9]

* * *

Back in the 1970s, the whole of Hayling's south shore was sandy: an essential part of the bucket and spade holidays that it supported and the associated seasonal infrastructure that created viable livelihoods for permanent residents. This was pre-package holidays, when Spain was still for the relatively wealthy and holiday camps drew city residents out to the coast. My family ran Merryland Amusements on Creek Road, the east end of the island, penny slots and a rock shop, grandma in the cafe and dad on the bingo. You've seen the rest—now try the best; pulling punters in with their ice creams and wind burns, sunrise till midnight Monday to Sunday, all summer. Saving for the winter and its dearth of work, ebb and flow. But the sand did what sand does and it followed the water out into the deep channels of the Solent. This would have kept happening and would have eventually progressed until the seafront was stripped bare, and so in the 1980s, the local council laid over 45,000 tonnes of protective shingle across the whole of the south shore. The sand beneath became a bed to part-memory, a door to the past that would open only at low tide.

The arcades and rock shops have dwindled down to a few now, along with the bucket and spade, but there is still sand to be found on Hayling, primarily on the seafront's west end: East Winner sand bar. East Winner is rippled and pitted and stretches one mile out when the tide is low. The more islanders I talk to, the more I'm advised to visit the sand bar at low tide and walk out onto the only unmarked part of the island. It's the most special place, I am told, on the low and turning tide; this is where time becomes flexible, bending and shifting with the pull of emotional attachment. The sand of both the East Winner bar and the beach as a whole is articulated as the "natural" landscape, lying partially hidden beneath what is therefore implied to be the "unnatural" bank of protective shingle. The fact that the shingle arrived and continues to be moulded by concerted human efforts (wearing hi vis vests, no less) seems to lend it a sense of artificiality, as if by being built by overt human action makes the pebbles themselves denatured; with a few notable exceptions, shingle beaches are not routinely described in the lyrical and wind-swept language that sandy beaches enjoy. Meanwhile, the sand particles that lie beneath Hayling's shingle are imbued with an aura of the "natural", that loaded word that speaks at once of purity, timelessness and authenticity. There are no red letter signs. No "Private Drive, Strictly No Turning", no "Trespassers Will Be Prosecuted". East Winner is the twice-a-day commons, free to all feet. We are welcome to walk on what was once sea, to tread the in-between and lean into the beauty of temporary, of constant change. So, of course, we do.

* * *

A hard and gravelly sand
dry at low water
& on it always
runs a
great
Tide[10]

* * *

In "Shifting Sands", an article focusing on an island in North Malé Atoll in the Maldivian archipelago, Uma Kothari and Alex Arnall set out to narrate "the multiple temporalities and rhythms of the movement of sand".[11] Their collaboration is rooted in the natural and human-made particularities on and around this unnamed place: a harbour and reclaimed land; the barrier reef and sand pumps; beaches and lagoons. Yet, although the article explores "the speed, pace and cadence of the passage of sand in, around and beyond" this one small island in the Maldives, the geographers simultaneously use that island to open up wider theoretical thinking about the complex relationship between islands and sand. Kothari and Arnall use this specific place to argue that, on islands, "localities where multiple forms of agency encounter and intersect, sand, with its multifarious movements and forms of stillness, haunts and enlivens the environment, the economy and society".[12]

In advocating "a greater sensibility towards, and understanding of, the movement of sand and its role in place-making", the authors make three wider contributions to the study of sand. First, Kothari and Arnall are attentive to the ways that "non-human agents, including sea, air, animals and plants, shape the temporality and spatiality of the movement of sand". Second, they endeavour to braid consideration of "the cultural and the material" by examining the "numerous multisensory and emotional responses" triggered by island sandscapes. Third, they attempt to bring together the radically different "temporal framings and perspectives" adopted by researchers— geomorphologists and ecologists, economists and socio-cultural theorists— for whom the substance of sand represents a shared object of scholarly attention.[13]

The authors, then, endeavour to understand sand through the intertwining of multiple disciplinary positions and conceptual frameworks. Informing much of the discussion, however, is a persistent preoccupation with process. Building upon Isla Forsyth's thinking on performance and the acknowledgement of the non-human, Kothari and Arnall introduce their article by referring to "the choreography of sand as a place-making process that occurs across different, interconnected temporalities".[14] Then, throughout the article, the authors consistently present the island as a sandscape that unfolds through the dynamic interaction of multifarious agents. The resultant effect is to suggest that: "Sand, a granular material, enables a narration of the island as always becoming".[15]

Hovering behind, above and beyond this always becomingness are two significant issues. The movement of sand, on this particular island, partly results from the economic imperative to satisfy the (mostly western) tourist's preconception of "a timeless, unchanging tropical landscape".[16] Sand is pumped and relocated to create the paradisiacal set on which tourism practices can unfold, while—behind the stage—the residents' landscapes are subjected to constant and insensitive modification. Alongside this, there is the desperately acute awareness of the Maldives as low-lying islands whose environmental vulnerability has led to an exodus of climate emergency refugees. The threat of catastrophic change adds an apocalyptic inflection to the understanding of the island sandscape as "always emergent, continuously recomposed by human and non-human agents".[17]

* * *

CBK Kitesurfing Website
Hayling Island is the birthplace of windsurfing and rightfuly [sic] *boasts some of the best kitesurfing conditions in the world, the East Winner sand bar which appears 4 hours either side of low tide creates a huge flat shallow lagoon for over a kilometre out to sea producing perfect learning conditions, which can be ridden in any wind direction.*[18]

* * *

Before travelling south, I had a fleeting conversation with a colleague who'd grown up in the area. When I asked her favourite place on Hayling, she said, without hesitation: "East Winner. It belonged to us locals. It's where I learnt to ride". Back in the office, I couldn't find this toponym on any of the paper maps laid out on my desk; nor could I find it on any of the digital maps open on my desktop. Eventually, though, I spotted it on the bottom of a sea-chart of Langstone and Chichester Harbours: a diagonal protrusion, in dusty yellow, off the south-west tip of the island. Clearly, in looking *within* the black-lined borders of Hayling Island itself, I had been searching in the wrong place.

In January 2014, the skeleton of a "100ft, 85-ton schooner"—lost in storms in 1865—was discovered at this site.[19] Marine archaeologists had to work against the clock in order to document the wreck. *The Ocean's* appearance, though, proved to be short-lived. As Julian Whitewright of the Maritime Archaeological Trust put it:

The wreck has now been swallowed up by the sands again – there are just a few frame tops sticking up above the sand. In one sense this is excellent as it is now fully protected again, although it means that studying it further is very difficult.[20]

For the archaeologists, East Winner conserves the nineteenth-century schooner; *The Ocean* remains safely buried beneath the sand.

* * *

"East Winner Bank Shipwreck: Archaeological Site Visit", Maritime Archaeology Trust.

Access to the site is limited to a period of around 1 ½ hours, straddling low water. This results in a reduced window of around 45 minutes when the water is at its lowest and conditions for working on the site are optimal. Even at that point, significant areas of the site continued to be underwater during the normal spring tide conditions of 0.6m (CD) during which the site was visited. Accordingly, successful work on the site required limited objectives that could be effectively completed within the access window.[21]

* * *

On the floor of Hayling Library, I sift through local maps and quickly find some that draw the eye to the spaces around, rather than within, the inverted T that is Hayling's perimeter. I unfurl a chart of Langstone and Chichester Harbours and blur my focus: yellow emerges as the dominant colour. In refocusing my vision, it becomes clear that white is used, on this particular map, to denote both land and shallow sea waters while yellow signifies the expansive sands off the Hampshire coast. Visually, this chart blurs the distinction between land and water; but, in contrast, a clear demarcation is made between sand and the environments that surround it. In this particular cartographic vision, the named topographies of Hampshire emerge as in-between spaces, and the sands themselves—Pilsey and Sinah, Mallard and Sword—emerge as locations in their own right. This chart of Langstone and Chichester Harbours functions as a mapping of sandscapes.

Sand, though, can confound cartographers. As I search through the charts, I start to notice an irregularity, an attempt to map time by mapping sand: some of the maps use the high tide mark as the definitive border of the island, while others allow the sea to give way to land, marking East Winner entirely in its temporary low tide state. According to the cartographers, it both is and is not. Sand and water, almost-land, and I wonder if that is part of its pull.

I look from map to map and wonder when they were made. I don't mean the year in which they were charted but, rather, what time of day. According to some, the coastline south of Hayling's Funland is clearly and unambiguously fixed. There are land, sea, and a bold black line separating the two. This, though, is a cartographical sleight of hand. The bay fills up and empties; the sand dismantles that illusory border. On many of the historic maps, the East Winner sand bar is submerged beneath the blankly blue space of Hayling Bay. The harbour chart, however, is different. Here, East Winner becomes the object of attention as the "always becoming" bar becomes fixed in both time and place.

I ask islanders about the sand bar and without exception their faces shine with the light of stories:

there was that time we walked right to the end, to the furthest tip, eyes on the horizon and the Isle of Wight and when we turned around again the tide had cut us off - we had to swim back, remember?

Auntie Anne smiles when she retells this. Swimming back fully clothed was the tiniest price to pay for a whole afternoon of almost walking on water, and the fact that the bank disappears each day makes it more precious. There's a special kind of love for a place that isn't. A transient land, a nobody owns it land, an unclaimed spit of sand open palm up, lifeline stretching from Langstone Harbour to the sea, heart line, love line, it's all here in the skin of your hands land. Both rough and soft at the same time, like the sand. Seaside fortune. Like those ripples, the steady pulse of the half-place, the sometimes, the twice-a-day liminal lick of land. Mental maps are drawn with emotion, and so they always include the sand.

* * *

Godfrey Baldacchino, "Re-Placing Materiality: A Western Anthropology of Sand"
[...] can one entertain memory and belonging without materiality? Is it not "things" which, become seeped in, and with, social memory in their production and consumption? Is it not materials which perform the past in their existence in the present?[22]

* * *

Havant Borough Council, "Havant Borough Townscape, Landscape and Seascape Character Assessment"
LCA38 Langstone Harbour mouth.
Relationship to Adjacent Character Areas
LCA39 South Coast Hayling Island: *The East Winner sand bank is a product of the coastal deposition at the harbour mouth, and at low tide this contrasts with the shallow sea of LCA39. Good intervisibility, although further to the east the gap of the harbour mouth becomes obscured as it visually blends in with the adjacent land masses.*[23]

* * *

Kothari and Arnall write about "the choreography of sand" as the place-making processes that occur across different interconnected temporalities—processes that are closely entwined with emotional and sensory reactions. Sandy place-making plays out on the Hayling seafront as a piece performed by both human and elemental dancers; the sea tugs and rolls the stage into a dynamic topography of shifting particulate while, further up the beach, waves of shingle are collected in dump trucks each autumn and driven along the seafront, returned to the unprotected eastern tip of the island—Greek drama on a Sisyphean shore. And if there is choreography, then there must of course be a choreographer. Lunar, human, gravitational. Close at hand there are the daily and seasonal tides that act on the sand and shingle: their dance partners the drivers, the trucks moving rock. Then the maritime traffic of Portsmouth dock, military vessels and container ships burning bunker fuel, the commerce that demands it, two clicks for a top from China, delivery in seven to twenty-one days. Carbon monoxide collects in the upper atmosphere, seas rise. The tide turns, sand hollows, music shifts. The people I speak to on Hayling build a relationship not only with the East Winner sand bar itself, but with its transience, both now and in a future where water changes everything. As plans emerge for more homes to be built on a sea-level, water-bitten island—bringing with them concrete-mixed sand—East Winner is out of reach of the developer's feet. And so it remains a common, for now.

* * *

"East Winner Bank Shipwreck: Archaeological Site Visit", Maritime Archaeology Trust
The tidal flow in the Langstone Harbour channel is dominated by the ebb tide when tidal rates can reach 1.5 knots (Bruce, 2008: 44-45). This has had the noted effect (SCOPAC: LT7) of flushing sediment seaward from the Langstone harbour channel to be deposited along the western side of the East Winner. The SCOPAC project also notes (O1) that there has been previous suggestion that the East Winner bank itself is partially fed through the westward movement of sand from the Chichester tidal delta. Although the mechanics of this are not proven, the overall sediment transport pathway within Hayling Bay is considered to be from east to west, allowing for deposition of material onto the East Winner (SCOPAC: O1) (p. 4).

* * *

He'd lived on the eastern side of the island for almost four years. He'd been resistant to the move at first: he was content in his own company and hadn't felt the need to be "sheltered". Eventually, though, he relented when his daughter pointed out that Sandy Point would be at the end of the road. Sand has always been his substance, and, here, he'd have a new beach to explore. Yet he struggled to feel at home. Having spent most of his life on the west of Hayling, he was used to the evening dazzle; but, here, it was all about mornings and the rising of the sun over Chichester Harbour. More than anything, he was used to stepping off land and heading out towards the beyond. He was used to the sandscape of East Winner.

It started not long after he moved. His daughter came down from London and brought a map of Hayling in the 1950s. She'd read an article in the paper about how maps can help with memory loss, so she went online and bought a map which might remind him of his childhood in West Town. It worked. He spent the rest of her visit deep inside himself, re-treading the lanes that he'd once walked to and from school. His fingers scored the memories back into place. Yet the map troubled him too. As his eye roamed to the bottom of the paper, he noticed that East Winner—the playground of his adolescence—was outlined but unnamed. Once his daughter had gone, he cut around the edges of East Winner and Blu-Tacked the remaining map onto the wall above his bed.

The next month he put in his orders. He wanted more maps and he wanted those maps to go beyond the 1950s. He also wanted different types of maps: sea charts and geological surveys; streets plans and cycle routes. He built up a cartographic library of Hayling: each sheet unfurled and attached to the wall; each sheet containing an East Winner-shaped hole. As his bedroom was only small, it didn't take long for the walls to be completely covered. He soon got to work, then, on the living room, the kitchen and the bathroom.

Yet, although Hayling was all around—both inside and outside his flat—he longed to *feel* place. One summer afternoon, he went across to the bureau and removed all of the East Winners that he'd been cutting out. Spreading them on the small kitchen table, he started chopping again. Once he'd cut the East Winner shapes into tiny pieces, he cut further still. He carried on cutting. He was determined to granulate the maps.

The pieces got everywhere. Later on that day, he found fragments in the sandwiches he had for tea. That evening, he discovered that thousands of microscopic pieces had congregated in the bottom of a Sainsbury's carrier bag. Then, when he took off his slippers and socks just before bed, the mapped East Winners began to gather in the spaces between his toes. He continued to feel them there as he turned off his reading lamp and listened for the sound of the sea.

* * *

Legend has it that Tennyson's famous poem, "Crossing the Bar", was written after a particularly rough crossing of the Solent as he returned home to the Isle of Wight, which lies south-southwest of East Winner. In the poem, Tennyson wishes:

And may there be no moaning of the bar,
When I put out to sea,

But such a tide as moving seems asleep,
Too full for sound and foam,
When that which drew from out the boundless deep
Turns again home.[24]

Tennyson completed "Crossing the Bar" three years before his death, and if his words are the poetry of the Solent waters, then every year on Hayling the sea becomes more poetic, swelling over the sand bar a little sooner, higher, fuller. With the rise of the saltwater comes an end, or at the very least a redrawing of maps—perhaps one in which East Winner is consigned to sandy memory. I ask friends and strangers about the bar and the surf catches sun in their eyes and glitters; I ask about the steady rise of the sea and their faces turn. It is coming. For the low lying, for a long time now, and everyone on the island knows it. There are no climate change deniers on Hayling. The sea continues to swallow the sand and there is disagreement about time: some guess at one hundred years, some two hundred before it takes the land. Some sooner. Nobody knows, really. No one but other islanders, other sea-level neighbours of sand arms that slip beneath the waves twice each day and with every storm: Kiribati, Palau, Fiji, Tuvalu, Micronesia, Cape Verde, Sarichef Island, the Maldives. Places that so often rely on the fickle economy of sand-supported tourism, places where the line between land and sea shifts. East Winner is still marked on most maps, but when will it dip below and become a shoal, something noted only on the nautical charts? When will the sand no longer be land-linked and drawn in pointillist curves, on our records at least? Right now, the bar is the protective barrier between the harbour and the sea, shielding Hayling's western shore from pollution, from simply being swept away. It protects the beach, and so protects tourism, which protects islanders. On one side are the warm shallows and on the other is a channel to Tennyson's boundless deep.

What is the future for life at sea level? When will the human choreography of sand and shingle prove too little?

Several locals tell me about the bells that are said to ring out in the bay, about how Hayling's sandy shore was once *so* much further out that not only have acres of farmland been lost over the centuries, but an entire church too. Although historic record does seem to show that Hayling was inundated during the fourteenth century, in particular, I can find no evidence of a lost church—only "Church Rocks" submerged in Hayling Bay.[25] But legends have a way of persisting, and the place of worship is collectively imagined to be able to withstand the onslaught of the sea; listen out on a clear still night and you'll hear the bells, I am told. I listen. Look on the old maps in the library and none of it is marked—no churchyard, no bell tower. All water. At some point our worlds pass from memory and fill the bottom of the deepest channels, the silt of before. I write about Hayling's present and future and the encroachment of the sea and I find myself slipping into the plural—"we", "our"—claiming my family inheritance of legends, an emotional connection to the island's sand despite only ever having known it, consciously, as a visitor. Are these stories the maps we must consult for guidance on our future, once-upon-a-times passed on, and down? Or do we look to the calm methodology of bureaucracy: the dry reporting commissioned by local councils and obscure branches of government, briefings read by fifteen people and the impression of control through the order of the official document. If bureaucracy is to be our guide, then we have a long history from which to draw.

* * *

"Coastal Management: Mapping of Littoral Cells", H R Wallingford Report SR328 (January 1993)
SUB-CELL 5a Selsey Bill to Portsmouth Harbour

COASTAL AUTHORITIES
District Councils: Chichester, Havant, Portsmouth
County Councils: West Sussex, Hampshire (England)

GENERAL
The coast is an eroding one and coast protection works have a knock on effect.

BEACHES
Shingle storm ridges over sand lower foreshores.

LITTORAL PROCESSES
A moderate westward drift due to southerly and easterly waves (predominant southwesterly waves diffracted towards north by the Isle of Wight). Rate of drift in recent years has been considerably reduced due to coast protection works and the diminution of beach material supply has been causing erosion from Selsey to Hayling. The drift is intercepted by harbour mouths and beach material is transported offshore onto tidal deltas by rapid ebb currents.[26]

* * *

In the remote language of official documentation, we again witness the chore-ography of sand: the diminution of beach material—diminution meaning not only to decrease but also, in musical terms, to repeat a theme of notes in one half or one quarter the length of those notes in the original theme. The music speeds up, and the choreographers ensure that the sand and shingle dancers keep pace with the accompaniment.

The report continues:

> Some evidence of an onshore supply of material along with frontage but supply is very small compared with "demand".[27]

Those quotation marks speak of euphemism. The "demand" of the sea (pull of the boundless deep) against the supply of the shoreline. The demand against which we are individually so ill-equipped to meet; the demand that we collectively stoke by feeding the engines of climate change, melting ice caps and drowning sand. Demand drives lives, it drives motion, migration. My family lived on the end of the island routinely described as most at risk of being overcome by the sea. Before they left to follow work as it departed, they sat in the pub on a spring tide and lifted their feet as the sea swept in through the front door, across the floor and then out again. "Everyone just laughed", Mum tells me, "and then we finished our drinks".

* * *

"Hayling Seafront Masterplan 2012", Havant Borough Council
> The Hayling Island coastline is at a significant risk from both flooding and coastal erosion. This risk is likely to increase considerably with projected climate change in subsequent years. The East Solent Management Plan aims to manage this threat by employing appropriate coastal protection techniques in strategic locations along the coast.[28]

* * *

Between 1941 and 1950, the aeronautical engineer and popular novelist, Nevil Shute, lived at Pond Head House: a large property in south-east Hayling. Over the years, local rumours have attached themselves to Shute's time on the island: stories of séances, held in Salt House, encouraged by a wartime government keen to establish contact with the Knights Templar who, so it goes, had once brought the Holy Grail onto Hayling.[29] Shute wrote prolifically during those years at Pond Head House. It was after emigrating to Australia, though, that Shute wrote his most famous work of fiction: the apocalyptic novel, *On the Beach* (1957), that documents the spread of deadly radiation from the Northern to the Southern Hemispheres following nuclear war. Focusing on four main characters located in and around Melbourne, "*On the Beach* finds its premise", according to Eleanor Smith, "in the spatial positioning" of that city "on the 'edge of the world'".[30] Melbourne, as the world's southernmost major city, is the last conurbation to be "desiccated" as a result of "the relentless, inescapable advance of the zone of radioactivity, removing all trace of human life from latitude after latitude on its way" towards Antarctica.[31]

Much of *On the Beach* focuses on the absurdly quotidian activities of characters who know—with varying degrees of acceptance—that time is short. The phrase, "on the beach", is a naval "slang expression normally, and originally, meaning retired from Service, but [...] sometimes used to describe an appointment to a shore establishment".[32] The title of Shute's novel, then, refers to the fact that Dwight Towers—the captain of an American submarine docked in Melbourne—finds himself spending more time on land than out at sea. As Smith points out, though, the beach also plays a significant narrative function, operating "as a persistent setting that [...] signifies the characters' denial of their fate".[33] The main actors sunbathe and swim, picnic and drink on the sandscapes of Victoria as catastrophe creeps towards them. As Smith also explains, the beach in the novel is symbolic as well as material in that it comes to represent "the liminal space" that the characters "occupy as they enter the final months of life". This trope of liminality was unambiguously visualised when, on the original cover of the text, the four main characters were depicted "on what Shute explicitly stipulated as the banks of the River Styx".[34] The novel, conforming to apocalyptic type, ends with the characters about to cross that threshold as they take their own lives rather than die from the effects of contamination.

In many ways, *On the Beach* is clearly a cultural product of its geopolitical time: a nightmarish response to the post-Hiroshima posturings of the Cold War era. Read sixty years after its publication, however, Shute's apocalyptic vision carries a different resonance in that it "provides a timely, frightening reflection of the crisis – and responsibility – reverberating through our own everyday lives".[35] This Anthropocenic resonance is deepened by returning to the quotation from T. S. Eliot's "The Hollow Men" (1925) that provides the epigraph to the novel:

> *This is the way the world ends*
> *This is the way the world ends*
> *This is the way the world ends*
> *Not with a bang but a whimper.*[36]

In the final weeks of Shute's irradiated world, his naval characters take to the water to peer at deserted cities and towns through a periscope, diligently recording radiation levels in the air above and noting the stillness of the land. They repeat, again and again, that there is nothing they can see—no bombsites, no damage—and fret about the quality of their reports given that they have only absence to log. In a conversation about their inability to comprehend that the end is coming, John, a civilian scientist, cuts in:

> "No imagination whatsoever," remarked the scientist. "It's the same with all you service people. 'That can't possibly happen to *me*'." He paused. "But it can. And it certainly will."[37]

On Hayling, we observe the sea from the land. We use our maps and reports and memories and histories and hearts to mark sand as ours, even if only temporarily, even as we see its diminution and hear the music speed up, and in doing so, we hold onto the sand bar through high tide and storm, and we try to defy the arrival of the deep.

* * *

Julian Hoffman, "Irreplaceable"
    We live in an age of diminution, thinning, disappearances. We live alongside shadows – ghosts of our own making. There is no easy way to convey the magnitude of loss currently underway in the natural world.[38]

Endnote

This piece of writing is an experiment in collaboration. To begin, our shared ambition—as a writer of short stories and as an academic—was to put creative non-fiction and critical writing together within the one textual space. Our hope was that the situating of these two forms side by side would lead to a thickening of the portrait of place. As the writing progressed, however, the boundaries between these two seemingly distinct modes began to break down: grains of literary critical thinking found their way into the sections designated as "creative"; particles of creativity seeped into the scholarly elements. Our habitual ways of writing about place, then, began to coalesce.

On reflection, therefore, we are no longer to assert with absolute certainty who was originally responsible for authoring which particular sections. More than that, we are no longer certain who was originally responsible for which ideas as, through the iterative processes of conversation and collaboration, the boundaries between our ways of thinking—as well as writing—about place have dissolved. It might seem surprising, then, that the chapter still contains the occasional use of the first person. We could have sought to have clarified such uses of the "I" by attributing authorship through paratextual notes. This, however, would have been an illusion and a denial of the complex aggregate that this work has become. Rather than imposing what would be a false division onto the narrative voice of this piece, we have instead taken not only our subject but our form and our approach from the ever-shifting ground of Hayling's liminal border, where the sand refuses to be neatly mapped.

Notes

1. Text from an information board erected by the Kitesports company, CBK Hayling Island.
2. Patrick Barkham, *Islander: A Journey Around Our Archipelago* (London: Granta, 2017).
3. John Kerrigan, *Archipelagic English: Literature, History, and Politics 1603–1707* (Oxford: Oxford University Press, 2008).
4. Jos Smith, "An Archipelagic Literature: Re-framing 'The New Nature Writing'", *Green Letters*, 17, no. 1 (2013): 5–15.
5. Smith, "An Archipelagic Literature", 2.
6. Stewart Williams, "Virtually Impossible: Deleuze and Derrida on the Political Problems of Islands (and Island Studies)", *Island Studies Journal*, 7, no. 2 (2012): 215–234 (215).
7. Sandra Dawson, "The Battle for Beachlands: Hayling Island and the Development of Coastal Leisure in Britain 1820–1960", *The International Journal of Regional and Local Studies*, 3, no. 1 (2007): 56–80.
8. Anon., "Hayling Island", in *A History of the County of Hampshire*, ed. William Page, Vol. 3 (London: Victoria County History, 1908), 129.
9. Havant Borough Council, "Hayling Island Seafront Regeneration Analysis and Feasibility Study" (2019), 91, accessed 9 August 2019, https://www.havant.gov.uk/sites/default/files/documents/Hayling%20Seafront%20Regen%20Study%20Reduced%20size.pdf.
10. Text taken from hand-drawn map from circa 1759 held in Hayling Library.
11. Uma Kothari and Alex Arnall, "Shifting Sands: The Rhythms and Temporalities of Island Sandscapes", *GeoForum*, 108 (2020): 305–314 (305).
12. Kothari and Arnall, "Shifting Sands", 306.
13. Ibid.
14. Ibid., 305.
15. Ibid., 306.
16. Ibid., 312.

17. Ibid., 309.
18. Anon., "Hayling Island", *CBK Kitesurfing*, accessed 9 August 2019, https://www.cbk-haylingisland.com/location/.
19. Anon., "Storms Reveal 1865 Wreck Off Hayling", *Portsmouth News* (15 July 2014), accessed 9 August 2019, https://www.portsmouth.co.uk/news/people/storms-reveal-1865-wreck-off-hayling-1-6178927.
20. Ibid.
21. Maritime Archaeology Trust, "East Winner Bank Shipwreck: Archaeological Site Visit" (May 2014), 5, accessed 15 January 2020, https://www.maritimearchaeologytrust.org/uploads/publications/MAT_EastWinnerBankShipwreck_May2014.pdf.
22. Godfrey Baldacchino, "Re-Placing Materiality: A Western Anthropology of Sand", *Annals of Tourism Research*, 37, no. 3 (2010): 763–778 (764).
23. Havant Borough Council, "Havant Borough Townscape, Landscape and Seascape Character Assessment" (February 2007), 26, accessed 9 August 2019, https://www.havant.gov.uk/sites/default/files/documents/Section%201.pdf.
24. Alfred Tennyson, "Crossing the Bar", in *The Poems of Tennyson*, ed. Christopher Ricks, Vol. 3, 2nd edn. (Harlow: Longman, 1987), 254.
25. Charles John Longcroft, *A Topographical Account of the Hundred of Bosmere* (London: John Russell Smith, 1857), 215.
26. J. M. Motyka and A. H. Brampton, "Coastal Management: Mapping of Littoral Cells", H R Wallingford Report SR 328 (January 1993), 41.
27. Ibid.
28. Havant Borough Council, "Hayling Seafront Masterplan 2012", accessed 9 August 2019, https://www.havant.gov.uk/sites/default/files/documents/Master%20Plan%20Update%202012.pdf.
29. See *The Secret Island of the Holy Grail*, dir. Robin Walton (Summerdale Productions, 2006).
30. Eleanor Smith, "The Poetics of Size: Rendering Apocalyptic Scale in Nevil Shute's *On the Beach* and Cormac McCarthy's *The Road*", *Colloquy: Text, Theory, Critique*, 35/36 (2018): 82–98 (86).
31. Paul Brians, *Nuclear Holocausts: Atomic War in Fiction 1895–1984* (Kent: Kent State University Press, 1987), 20.
32. Anon., Royal Navy Diction and Slang, accessed 9 August 2019, https://www.hmsrichmond.org/dict_b.htm.
33. Smith, "The Poetics of Size", 89.
34. Ibid.
35. Ibid., 95.
36. T. S. Eliot, "The Hollow Men", in *Collected Poems 1909–1962* (London: Faber & Faber, 2002), 77–82 (82).
37. Nevil Shute, *On the Beach* (London: Pan Books, 1973), 79.
38. Julian Hoffman, *Irreplaceable: The Fight to Save Our Wild Places* (London: Hamish Hamilton, 2019), 97.

14

Drifting in a Cemetery of Sandscapes

Julian Brigstocke

The seaside town of Deal is a small, popular, prosperous tourist destination in Kent, UK. Once an important port, it is now a well-regarded tourist destination. Holiday makers enjoy the Georgian architecture, the 300-metre concrete-clad steel pier, the winding streets, and the sandy beach. This beach is under threat. Recently, a £10 million project commenced with the aim of defending the coastline, building a 410-metre concrete sea wall, amongst other things, to reduce the annual risk of flooding. East Kent suffers from serious coastal erosion and flood risk. Major beach works continually struggle to keep the coastal beaches intact. Known as "holding the line", beach nourishment involves maintaining beaches and returning the sand when it drifts elsewhere. This is not only important because the beach is a vital amenity and tourist attraction for the local community, but also because it acts as a natural flood defence. During winter storms, sand and shingle drift northwards. Work is required to bring lost sand to eroded parts of the beach before it disappears forever. Maintaining this sandscape, fighting against the natural drift of the sand, is a difficult and costly endeavour.

* * *

In this chapter, we too shall drift, as we explore how sand links up disparate places, politics, and forms of power. We drift from British seaside towns, to the ocean floor, into African winds, towards deep time and colonial history.

J. Brigstocke (✉)
Cardiff University, Cardiff, Wales, UK

© The Author(s) 2020
J. Carruthers and N. Dakkak (eds.), *Sandscapes*,
https://doi.org/10.1007/978-3-030-44780-9_14

Might thinking and moving with sand, in drifts and swirls and slow accretions, help us find our way towards a planetary ethic that welcomes the body of the Earth into our experience of self and responsibility?[1] Thinking with sand reminds me that I am not only a biological, social or ecological being; I also have a geological self which is entangled with the residue of thousands of years of grinding and wearing and weathering. Thinking with sand, we form connections between disparate times, places, and politics.

* * *

A couple of miles out into the North Sea, we find a ten-mile area of gently drifting sand banks, made up of sand and coarse sediment about twenty-five metres deep. These are known as Goodwin Sands. The sands continually change with the ebb and flow of the tide and the movements of the sea. Continually on the move, the sands have proved hazardous to ships throughout the ages, and from time to time, shipwrecks become visible on the sands, before disappearing again. Thousands of shipwrecks lie on the sea floor. Over 2000 lives were lost to the sands on one single night during the Great Storm of 1703. The sands are home to a wide range of marine life, including a large harbour seal and grey seal population. They are also foraging grounds for many bird species. In May 2019, the Sands were designated a Marine Conservation Zone, mandating the conservation or replenishment of features including subtidal sand and subtidal coarse sediment (pebbles). The water around the Sands includes areas of sand and pebbles that nurture very high biodiversity. It contains Ross worm reefs and blue mussel beds. Circalittoral rock shelters bryooans, pink sea fans, cup corals, anemones, soft corals, sponges, sea squirts, red algae.[2] Thornback rays deposit their uniquely frilly mermaid purses.

* * *

Sand has strong associations with leisure: beaches, the seaside, sandcastles. It is wonderful to play with because of the curious nature of granular materials: sand is a solid that behaves and flows a lot like a liquid. Yet if we think a little harder about the role of sand in our lives, we soon start to notice sandscapes everywhere. Sand has myriad human uses and is the third most heavily consumed resource on earth, after water and air. Sand immediately connects us to global processes: the construction industry, fossil fuels, transport, cities, glass, manufacturing. This has caused a sustainability crisis: sand is a non-renewable resource (it is created on geological time scales), and the planet is running out of construction sand.[3]

* * *

Much of the material in Goodwin Sands will soon participate in a form of drift that is rather more abrupt than its twice-daily dance with the moon through tidal flows. In 2018, and then a 2019 judicial review, the Port of Dover was granted permission to dredge three million tonnes of sand from Goodwin Sands to provide aggregate for land reclamation. A ship called a suction dredger will drag two large heads behind it on the seabed, sucking up material from the seabed like a giant vacuum cleaner. Anything in the path of the heads on the seabed will be destroyed. Sand will be piped up to the surface and loaded onto hoppers, and the dredger will then sail to Dover to unload its cargo.

Usually, marine dredging creates much less controversy than land-based sand mining: there are no neighbours to make a fuss. However, the removal of sand from Goodwin Sands provoked furious opposition. A few weeks after the Sands became a legally protected Marine Conservation Zone, a judicial review granted permission to dredge the zone. Environmental activists insist the unearthing of the seabed will destroy the spawning and nursery grounds of many fish, as well as the habitat of the sand eel, which is vital to the ecosystem of the area. They say there is also a risk that the dredging will shake up the sand and increase the turbidity of the water, making the waters uninhabitable for many fish and asphyxiating fish eggs. The colony of nearly 500 seals could also be disturbed by the noise and vibration of the dredgers, as well as seeing a reduction in their food supply. Yet the official report on the implications of dredging observes that the sandbank lies within a naturally dynamic environment, and that disturbance to the subtidal sand will be temporary and spatially limited.[4] The wildlife affected by the dredging, it goes on, should recover within "several years" after the dredging ceases.

* * *

Whenever we encounter a sandscape, we touch antiquity. Sand is a material that is mysterious, ancient, and infinitely varied. Any grain of sand is the result of an unimaginably long geological process of wearing, weathering, and decay. A sand granule links us to the beginning of life or to the beginnings of the earth itself.[5] A rock in the mountains is slowly ground down by the insistent beating of water, air, ice or rock. Eventually, it breaks up, its minerals wash away, and it decays into the smallest of granules. Sand is the fate of rock: ground down, worn away, and always on the move, however slowly, following the flows of rivers, streams, glaciers, storms, winds, and oceans. Each grain of sand is unique and links us to processes that go back to the origins of the world.

When we speak of sand, we do not refer to a specific substance, but a scale. In its composition, sand is mixed: it is a jumble of ground-up rock and organic matter such as shells and bones. Sand is impossible to define, other than as a material of a particular size: generally between 0.0625 mm and 2 mm. Any bigger, and we call it gravel; any smaller, and it is silt. Yet sand is not just a mixture of substances; it is also a mixture of scales. Sand invites us to jump across wildly different levels: the micro-scale of individual granules and their specific properties, as well as the spaces between them—the spaces where most of the life on a beach resides—but also the macro-scale of global drifts, environments, ecosystems, and the history of the Anthropocene, the geological epoch in which humanity permanently transformed the Earth's crust. Similarly, sand invites us into the temporal scale of momentary flux and change—the sands of time, the hourglass—but also the twice-daily ebb and flow of the tide, the slow drifts of sand across the years, the gradual weathering of rocks over thousands of years. Sand brings to mind the long history of human folly—the sands of Shelley's *Ozymandias*—and the vastest scales of geological time. Sand lures us into exploring multiple scales and disparate geographies.

Above all, sand is drift. Thinking with sand—walking upon sand, touching it, looking at it, smelling it—opens up questions about origins, metamorphoses, processes, decay, death, and the nature of time passing. Drift is the detritus of unconsolidated matter that lies on top of strata of consolidated, stable bedrock.[6] In geological terms, drift is very young, mostly formed within the last 2.6 million years. Drift is primarily sand, gravel, silt, and clay, as well as soil and peat. It occurs in thin, discontinuous patches. It drifts around the surface of the earth, generally moving downhill towards the sea, and then with the ocean tides and currents. It supports life as we know it. For this reason, drift has become the focus of environmental concerns, as these discontinuous patches of unconsolidated matter are concreted over, built upon, farmed, polluted, washed away, drained, excavated, mined, and despoiled.

Sand drifts naturally in three main ways: surface creep (rolling along the ground); suspension (drifting in a fluid medium, either air or water); and saltation (moving in hops and leaps).[7] Of these modes of transport, saltation is the most important: 75% of sand moves in this way. In saltation, a particle is lifted by the wind or water from the surface, is accelerated by the fluid, and then pulled downwards by gravity. Saltation is a kind of hopping motion, a movement by small leaps and jumps. In saltation, there is a continual dialogue between ground and the fluid medium; it works through a dynamic play between solid and fluid bodies. What's more, if the particle attains enough speed, it can eject or bounce other particles into saltation, banging sand grains and kicking them into the air. As this ballistic process rapidly gains momentum, a cloud of sand emerges, composed of sand grains leaping and jumping, kicking off other grains as they land. Such saltation bombardment creates chaotic, turbulent flows of matter such as dust storms and avalanches.

Might our forms of thinking and writing also learn to move through the hops, jumps, and leaps of saltation?

* * *

Sand is almost indestructible. It is what survives after the incessant grind of thousands of years of erosion has worn everything else away. Sand—mixture, multiplicity, difference—survives when everything else has been pulverised.

* * *

It is not only marine life that the Goodwin Sands shelter. They are home to another kind of archive, of thousands of ships that have foundered on the treacherous sands, as well as dozens of aircraft that came down during the Second World War. Initial surveys using side-scan sonar did not record any sites of archaeological interest in the proposed dredge area. After opposition groups funded a new survey using a magnetometer (a kind of metal detector), 243 sites of interest were identified, including five large objects, one of which has been subsequently explored by amateur divers and found to be an aircraft. However, archaeologists claim that this improved survey may still have missed any older wooden ships, as well as aluminium aircraft, since neither of these would be picked up by a metal detector. In fact, they claim that one of the most important maritime archaeological areas in England is set to be destroyed by the dredgers.[8] The first sign of discovering fragile archaeological material is when smashed-up pieces of it are brought up by the dredger. They point out that when a similar survey was approved for a dredge site in the waters off London Gateway in 2010, the survey failed to spot very rare archaeological finds which were subsequently destroyed.[9]

* * *

The material dredged from Goodwin Sands will be used as landfill for a vast construction project in the Port of Dover. This project aims to regenerate the struggling town. The Dover Western Docks Revival will deliver a new commercial waterfront with shops, bars, cafes, and restaurants; a relocated and redeveloped cargo business, including a new cargo terminal and distribution centre; and expansion of the Eastern Docks for ferry traffic.[10] This £200 million project will create a "port-centric logistics hub" that requires a large land reclamation project, with two marinas (Tidal Basin and Granville Dock), along with the Wick Channel, being transformed from sea into concrete.[11] Here, sand forms the vital material infrastructure for economic hopes, plans, and realities.

Since the 2016 referendum that voted for Britain to leave the European Union, it has been widely rumoured that much of the new development is being earmarked for a new lorry park, in case additional border checks for freight arriving from the European Union lead to massive logjams. In January 2020, a month after the port opened the new cargo terminal intended to equip the area for future import/export needs, Dover Port announced a major efficiency review that would lead to an unspecified number of redundancies.[12]

* * *

Campaigners against the dredging of Goodwin Sands also worry about the impact of dredging the sands on Kent's serious coastal erosion problems. Current models, they argue, are simply not accurate enough to be able to predict with any certainty what the removal of vast quantities of sand from the sea floor will do to the nearby coastline.

The sandscape of Deal beach exists in close connection to many other local sandscapes: the concrete sea wall and pier; the continual labour of beach nourishment; the natural flood defence of Goodwin Sands; the removal of sand from the sea to create concrete industrial sandscapes. Sandscapes are everywhere. But the politics of how and where sand is extracted from, and for what uses, are becoming ever more troubling. Sand is not an infinite resource. Whenever we build something, the material must come from somewhere. And we cannot always predict the effects of doing so.

* * *

Devonport, Plymouth, on the south-west coast of Britain, is the biggest naval base in Western Europe. In 1915, at the time of the Great North Yard extension, it was one of the biggest in the world. We find here, in concrete form, the sand whose disappearance has been widely blamed for the destruction of a local fishing village. In 1897, to procure materials to extend the naval base, the Board of Trade licensed the removal of material from the intertidal zone of Hallsands beach. As much as 660,000 tonnes of sand and gravel were dredged. Soon, local villagers started complaining that the dredging was causing their beaches to disappear. By 1902, the beach level in Hallsands had been lowered by 3 metres, and by 1904, it had lowered by 6 metres. By this point, 97% of the beach volume had gone. In 1900 and 1902, storms undermined sea walls and eroded sand and gravel from the rock ravines lying behind them. It had been assumed that any lost shingle would be replaced naturally by fresh material. However, the shingle was in fact deposited thousands of years ago during the last ice age and showed no signs of returning. With its natural protection destroyed, the village was highly vulnerable, and on 26 January 1917, a north-easterly gale, causing waves over twelve metres high, combined with a high tide to destroy the village and to lower the beach by another two metres.[13] The entire village, housing 127 people, was lost to the sea in one stormy night, though miraculously everyone was evacuated safely and there was no loss of life.

Two houses survived, now used as holiday cottages. In May 2012, a ten metre, two hundred tonne section of coastal cliff collapsed. Inhabitants of the two remaining houses were successfully evacuated.

Hallsands is now used by opponents of dredging the Goodwin Sands as a warning of the perils of interfering with complex tidal flows that are not fully understood or modelled.

* * *

Sand is never still. It drifts from one place to another. It flows in water, land, and air, and it affects the flows of water and air around it. Humans channel its flows, through the application of volcanic levels of heat, or by nurturing chemical reactions. In this way, sand is transformed into materials as varied as concrete, glass, cement, asphalt, microchips, and much else.[14] Sandscapes are not only found in beaches and deserts: we see sandscapes in landscapes populated by concrete walls, lorry parks, ports, container terminals, buildings, roads, windows, computers, and smart cities. I see sandscapes under my feet, in the Devonian era sandstone that is common in south-west England where I live. (Sandstone is one way in which the geological cycle turns full circle: a layer of sedimented sand is compacted by overlying deposits, and eventually cements together, with the help of minerals such as silica or calcium carbonate, into stone.) I see sandscapes in the air that transports sand across the world; in the rivers and oceans; in every building and road in the city. Like air and water, sand is everywhere.

Modern urban life has a sandy substrate. Its grounds are perpetually drifting and crumbling, repeatedly claimed, reclaimed, concretised, and destabilised. Even in concrete, where fluid sand is consolidated into a stone-like substance, the appearance of permanence is illusory: most concrete lasts only for decades, before being sent to landfill. Looking at sandscapes in this way, we rub up against the abrasive, sandy skin of modern life. Might we learn to attune ourselves to sand's devious drifts and granular flows? Might we refuse the demand to find the solid ground, and follow the teasing dance of unconsolidated ground? Doing so, perhaps, might help us find our way to an ethos that summons new kinds of reclamation: reclamations of identity, minerality, multiplicity, memory, time, and truth.

* * *

Sand often moves slowly, but it always gets on the move eventually. It can travel thousands of miles, whether with rivers, oceans, ice or wind. Take, for example, a grain of sand from a sample taken in Plymouth.[15] The sample was collected after a volume of Saharan dust was deposited on the UK mainland. Dust can travel suspended in the air for years, and frequently travels thousands of miles. For example, dust blown over from the Sahara plays a vital role in replenishing the nutrients that keep the Amazon rainforest alive.[16] Sand grains, however, which are larger than dust particles, should not be able to travel in the wind for any significant distance: gravity ought to win out before long. Physicists' models of granular transport only allow larger sand particles to travel for a few metres at a time, through saltation. Yet, on close examination, there is no doubt that our grain of sand has travelled, suspended in the air, all the way from the Sahara. How did it arrive here?

Each grain of sand is an archive, a unique sandscape in itself. If you observe a grain of sand through an electron microscope, you can read a whole historical geography from it. Most obviously, you can see what kind of rock it is made of, which will give you a clue about where it came from. We learn even more from its surface features. As sand travels through rivers, sea, ice, and wind, it is buffeted around in distinctive ways that imprint themselves on the particle's shape: how rounded or angular it is, how pockmarked it is, what kind of abrasions and ridges it has, and what kind of chemical weathering is visible. As sand comes to rest in a dune, a seabed or in the soil in your garden, different processes will continue to alter it. Our grain of sand found in Plymouth has clear hallmarks of having come from a very dry, arid, environment with salty winds: it has a dulled surface, creases, and pockmarks. We do not fully understand how these giant particles flew through the air for thousands of miles. Yet the evidence that large numbers of sand particles do fly across the globe is clear. Our sandscapes link us to distant places through unexplained mechanisms and to geological processes that occurred over millions of years on the other side of the world. The movements of sand remain a mystery.

* * *

Viewing the world as sandscape, we see it as life ground down: an incessant, inexorable process of erosion and weathering. We see what remains of the earth's crust from long erosions over thousands of years. This sandy residue, writes Italo Calvino, is "both the ultimate substance of the world and the negation of its luxuriant and multiform appearance".[17] The fate of all monuments, all stone, is to be ground down, sooner or later, into dust. Viewing the world as sandscape, we see a "cemetery of landscapes reduced to a desert". But, asks Calvino, might we still find, in sand, a foundation and model?

* * *

A group of environmental activists known as Reclaim the Power break into a quarry in Cheshire, owned by a company called Sibelco, the world's largest producer of sand for hydraulic fracturing (fracking). Sibelco mines frac sand in countries including the USA, Belgium, France, Russia, and the UK. The protestors scale scaffolding and put up banners saying "Sibelco, Stop Supplying Oil + Gas Firms" and warn them: "We'll be back". Campaigners are determined to alert the public to the fact that, despite a supposed "suspension" of fracking announced by the UK government in 2019, many fracking activities are not included in this supposed suspension.[18] Their previous actions have included occupying convoys of lorries delivering drilling equipment, delaying them for several days, and successfully persuading the company to dissociate itself from fracking firms.

Fracking requires vast quantities of sand. It fractures rock by shooting a high-pressure mix of water, chemicals, and sand into a well bore. Drillers then shatter the surrounding shale, creating a network of cracks that the gas can flow through. The sand is vital for keeping the cracks open, stopping the pressure of the surrounding rock from closing them up again. This requires a very specialised kind of sand; it must be at least 95% pure quartz, so that it can withstand very high pressure. The grains must all be small and rounded and therefore probably extremely old, not like the angular shape found in more youthful sand. Sand deposits combining this purity with this grain shape are very rare. In the USA, Wisconsin has deposits of the appropriate sand, known as "northern white", and in only a few years became devastated by dozens of vast sand mines, blowing up the sandstone hills and harvesting the sand. Wisconsin produced 25 million tons of frac sand in 2014, and this activity ruined landscapes, destroyed farmland, poisoned groundwater, and led to chronic lung disease. By 2019, many of the mines had shut down again, with the frac sand industry rapidly relocating to Texas. In the UK, frac sand mining is in its infancy, and protestors are determined to ensure that areas such as Cheshire and Norfolk, which hold frac sand deposits, do not suffer the same fate as Wisconsin. They are not only fighting to keep the oil and gas in the ground, but also the sand. The presence of frac sand bears omens of a devastated landscape.

The Congleton and Chelford Sands are a key site of frac sand. Sibelco has two sand mines in the area, supplying Cuadrilla, a controversial UK fracking company. Sibelco is planning a third mine, which has received significant public opposition. Sibelco's study showed that there are 3.3 million tons of Devensian-era silica sand at the site. If fracking were to take hold in Lancashire, then it is possible that many more sand mines would be needed. "This is not about one quarry", the campaigners worry,

> but the potential creation of what would effectively be a whole new industry which would chew up large areas of Cheshire […] Resisting the activities of frac sand companies like Sibelco is as important as opposing fracking companies like Cuadrilla. Communities inside and outside Cheshire need to work together if they do not want their local areas to be trashed.[19]

* * *

I visit a conventional sand quarry in the Cotswolds, close to my home in Bristol. From the road, the digging site is almost invisible, carefully hidden through landscaping. It is a relatively small quarry, producing construction sand for local businesses within thirty miles or so. Sand is heavy and therefore expensive to transport, so conveying it a long way is rarely cost-effective. The site manager shows me around. Diggers perform an intricate, hypnotic dance, forming an efficient circuit from the digging site to hoppers which pour the sand onto a conveyor belt running around the entire site. Digging is skilled work: if the driver gets it wrong, the digger may get stuck in the sand, or it may topple over, or it may pick up the wrong kind of sand: the wrong grade (size), for example, or the wrong composition. The conveyor belt brings the sand to the shaker deck: a huge metal box, shuddering like a giant washing-machine drum, which sorts the sand granules in a giant centrifuge. A tangential feed of quantities of water drives a rotational motion that causes relative movement of materials suspended in the fluid, separating and sorting the materials.

Whenever we dig into the ground, we unearth another archive. Digging here has been interrupted by numerous geological and paleontological finds: septarian nodules (about 60 million years old); ammonites (65–240 million years old); freshwater crocodiles (up to 65 million years old); steppe mammoths (a mere 360,000–600,000 years old). The owner of the quarry is excited by the finds, even though they delay digging. He is a geologist by training and enjoys talking about all the discoveries he has made in the land here. He knows better than anyone that digging for sand carries us once again into deep time and the origins of humanity. Keen to keep good relations with the local community, he has given findings such as a complete mammoth fossil to local tourist sites.

Sense of place is important. One profitable area of the business comes from selling some of the aggregate, at an enormous markup, as ornamental gravel for people to use in their gardens. Owners of picture-postcard Cotswold cottages, he explains, are keen to have local stone of exactly the right colour to fit in with the local oolitic limestone landscape. He is proud that his produce is considered so aesthetically pleasing, even if he is baffled by the prices people will pay for it.

Wherever you find sand, you often find water bubbling up. Sand and many types of sandstone are porous, so they provide a natural path for water. At this site, when the digging has finished, the pits are converted into a nature reserve, with large lakes, islands, as well as fields and agricultural land. The water and islands provide an attractive habitat for birds and other wildlife: herons, wading birds, sand martins, and Roman snails. This restitution of the land is an important part of the lifecycle of the sandscape. We walk around parts of the site that are being restored. One area of land, previously a field, is now scrub wilderness. An ecosystem is slowly emerging of its own accord, and we see bulrush, willow, and silver birch growing. Effectively, the manager tells me, we borrow the land for a few years and give it back to them better than it was before. He works endlessly to build good relations with the local community: he is hoping to receive planning permission to extend the site and to keep on digging for a few more years. But the planning process, he tells me, is getting harder and harder. Local residents are reluctant to agree to industrial work happening in their neighbourhood. They complain about the noise of the trucks. And yet our demands for sand are always increasing: building just one average-sized house requires 60 tonnes of aggregate. If we include the associated infrastructure, it can be as high as 400 tonnes. It is as if people don't understand that sand comes from somewhere, he tells me. If we can't dig up the sand, how can we build their roads and houses? It is as if people expect our building materials to come from nothing; as if sand comes out of thin air and does not have a place, a history, and a story to tell.

* * *

The myth of gaining land from nowhere saw its most violent expression in colonialism, which, by refusing to recognise indigenous inhabitants as people, could see their land to be empty and open to discovery and colonisation. Colonialism, that most devastating form of drift, has its own roles to play in the global archives of sand. The environmental effects of sand mining are severe, changing rivers and ecosystems, increasing suspended sediments and causing erosion. Rich, sand-hungry places such as China, Singapore, Hong Kong, Dubai, and Qatar now import as much sand as possible from nearby countries including Vietnam, Cambodia, the Philippines, Malaysia, and the Maldives. Effectively, they export the devastating effects of sand mining to poorer countries. The Mekong Delta has been mined so extensively for sand that the sustainability of the world's third largest delta, Southeast Asia's most important "food basket", is under threat, due to a combination of sand mining, dam-building, and groundwater extraction.[20] Shoreline erosion is already displacing coastal populations, and peasants' land is crumbling into the river as the banks subside. In Indonesia, at least two dozen islands have disappeared entirely, due to illegal sand mining. These countries have now banned exports of sand, but illegal exports of sand thrive—sand is very difficult to trace. In India, sand mafias operate in almost every one of India's four hundred rivers, resulting in large (but unknown) numbers of deaths and other violence. Sand mining has been blamed for the severity of the flash floods in the province of Uttarakhand in 2013, in which approximately 6000 people died and 110,000 had to be evacuated by military and paramilitary forces. Widespread criticism was voiced that unrestricted sand mining from the rivers had exacerbated the disaster by modifying water flows. Sand mining was briefly halted, but resumed within a few months. Sand mining has been blamed again for the severity of the 2018 floods in Kerala, the worst in nearly a century, killing at least 458 people.

* * *

Conflicts over sand have become ever more acute with the construction boom of recent decades. In British Hong Kong, however, we find evidence of conflicts over sand that date back as far as the early twentieth century, when the colony was in its infancy, only a few decades after the "barren island" had been ceded to the British Empire after China's defeat in the Opium Wars, during which the British Empire fought to protect their right to sell opium in China, in the name of "free trade".

In the Hong Kong government records office, you can order an archive titled "Hong Kong Colonial Sand Monopoly". Sifting through the archive, I encounter a fragmentary history of the dispossession of indigenous villagers by colonial sand digging. The colonisers, needing ever-increasing amounts of construction sand to keep up with the population boom, dug up any beach sand they could find. Without their natural flood protection, indigenous villagers found their paddy fields engulfed with sea water. Colonial sand extraction deprived indigenous villagers of their livelihoods, leaving them destitute. The earliest record I find is in 1906:

> Petitioner's families have lived in the Kau Wa King Village for more than a hundred years. They own a piece of taxed sandy land handed down to them by their ancestors. When their village was included in the British Territory in 1899 His Excellency the Governor on application by petitioners, graciously allowed them to hold that piece of land which the boundaries are clearly marked out in a plan. Since the year before last some boatmen in lighters have dug sand from within the boundaries of petitioners' land and they also destroyed the boundary stones fixed by petitioners. The said piece of land protects petitioners' field from inundation in case of bad weather. Petitioners hope that although they dare not stop this practice themselves, the law can punish those dishonest boatmen who profit themselves at the expense of others. Petitioners therefore pray Your Honour to order some policemen to investigate this practice of the boatmen.[21]

Incidents such as these proliferated. By the 1930s, villagers were piling stones in the sea to stop junks from getting near their beaches; if they succeeded, the villagers fought them off with guns. Elsewhere in the colonial bureaucracy, meanwhile, administrators became increasingly upset about the disappearance of sand from local swimming beaches. The government, realising the importance of sand as a strategic resource to ensure Hong Kong's prosperity, established a Sand Monopoly to regulate the extraction of sand and to keep every grain of sand under the ownership and strict control of the British Empire. Villagers who had their land dug up from under them were given short shrift. One administrator merely noted: "It seems thoroughly bad policy to take any action which will hamper building in the colony or increase the price thereof". Until sand supply was returned to the market (when Hong Kong finally ran out of sand and became reliant on importing it from elsewhere), the Sand Monopoly formed a small but vital pillar of the colonial administration. Without a continuous supply of sand, the colony could not exist. When supplies were delayed by even a couple of hours, it created a minor crisis in the city.

* * *

Our thought is not ours alone; it is made of the things that assemble our lives and bodies, as well as those things that make our worlds thinkable. Sand thinks through me and, to the extent that modern self is already sandy, vitrified, concretised, cemented, and microchipped, thinking with sand invites me to pay closer attention to those granular materials that think in, through, and with me. In its encounters with air, water, fire, tricalcium silicate, and many other substances, sand composes human societies and ecologies.

Sand drifts across spaces, times, and scales. It is equally at home in land, water, and air. It forms an archive of pulverised stones and fragmented stories. It exists in sandscapes that are as varied and mixed as sand itself. Learning to care for sand, perhaps, may encourage a thinking from the shoreline: a tidal poetics formed from alluvial deposits of sand, silt, and mud.[22] Following the drifts of sand, however, reminds us that geology too is a mode of accumulation or dispossession, depending on what side of the line you fall on.[23] Sandscapes are often sites of play, joy, and leisure. But they are never innocent. They harbour infinitely varied and fragmented stories.

Notes

1. See Bronislaw Szerszynski, "Drift as a Planetary Phenomenon," *Performance Research* 23, no. 7 (2018): 136–144.
2. Department for Environment, Food and Rural Affairs, "Goodwin Sands Marine Conservation Zone," accessed 31 May 2019, https://assets.publishing.service.gov.uk/government/uploads/system/uploads/attachment_data/file/805463/mcz-goodwin-sands-2019.pdf.
3. United Nations Environment Programme, "Sand and Sustainability: Finding New Solutions for Environmental Governance of Global Sand Resources," Geneva: United Nations, 2019, https://unepgrid.ch/sand/Sand_and_sustainability_UNEP_2019.pdf.
4. Marine Management Organisation, "Marine Conservation Zone (MCZ) Stage 1 Assessment: Aggregate Dredging at Goodwin Sands (Area 521)" 2018, https://assets.publishing.service.gov.uk/government/uploads/system/uploads/attachment_data/file/729818/20180725_-_Goodwin_Sands_pMCZ_Stage_1_Assessment.pdf.
5. See Rachel Carson, *The Edge of the Sea* (Boston: Houghton Mifflin, 1955), 123.
6. See Deborah Dixon, "The Perturbations of Drift in a Stratified World," *Performance Research* 23, no. 7 (2018): 130–135.
7. See Michael Welland, *Sand: A Journey Through Science and the Imagination* (Oxford: Oxford University Press, 2009), 150.

8. Joint Nautical Archaeology Policy Committee, Statement to Goodwin Sands SOS. August 2017, accessed 24 January 2020, https://goodwinsandssos.org/.

9. Peter Holt, "The Suitability of Pre-Disturbance Geophysical Surveys for Underwater Cultural Heritage in England," *3H Consulting* (28 March 2017), http://www.3hconsulting.com/Downloads/2017_3H_MarineGeoUCHProblems.pdf.

10. Port of Dover, "Delivering the Vision: Transforming Dover's Waterfront for Future Generations", accessed 24 January 2020, http://www.doverport.co.uk/port/about/dwdr/.

11. Port of Dover, "Dover Western Docks Revival (DWDR) Newsletter" (Autumn 2019), http://www.doverport.co.uk/downloads/DWDR_Newsletter_Autumn_2019.pdf.

12. Beth Robson, "Port of Dover Redundancies: Number of Job Losses Unknown as MP Natalie Elphicke Calls for Urgent Meeting", *Kent Online* (8 January 2020), https://www.kentonline.co.uk/dover/news/redundancies-announced-at-port-of-dover-219732/.

13. V. J. May and J. D. Hansom, *Coastal Geomorphology of Great Britain. Geological Conservation Review Series 28* (Peterborough: Joint Nature Conservation Committee, 2013).

14. See Vince Beiser, *The World in a Grain: The Story of Sand and How It Transformed Civilization* (New York: Riverhead Books, 2019).

15. N. J. Middleton, P. R. Betzer, and P. A. Bull, "Long-Range Transport of 'Giant' Aeolian Quartz Grains: Linkage with Discrete Sedimentary Sources and Implications for Protective Particle Transfer", *Marine Geology* 177, no. 3 (2001): 411–417.

16. Hongbin Yu, Mian Chin, Tianle Yuan, Huisheng Bian, Lorraine A. Remer, Joseph M. Prospero, Ali Omar, David Winker, Yuekui Yang, Yan Zhang, Zhibo Zhang, and Chun Zhao, "The Fertilizing Role of African Dust in the Amazon Rainforest: A First Multiyear Assessment Based on Data from Cloud-Aerosol Lidar and Infrared Pathfinder Satellite Observations," *Geophysical Research Letters* 42, no. 6 (2015).

17. Italo Calvino, *Collection of Sand*, trans. Martin McLaughlin (London: Penguin, 2014), 22, 26.

18. Frack Off, "Fracking 'Pause' Does Not Include Shale Oil in Weald (Sussex/Surrey)" (4 November 2019), accessed 24 January 2020, https://frack-off.org.uk/social-media-post/sussex-surrey-ot-covered-by-fracking-pause/.

19. Frack Off, "Why Does Frack Sand Mining Threaten Cheshire & East Anglia?", accessed 24 January 2020, https://frack-off.org.uk/faq/why-does-frack-sand-mining-threaten-cheshire/.

20. Edward J. Anthony, Guillaume Brunier, Manon Besset, Marc Goichot, Philippe Dussouillez, and Van Lap Nguyen, "Linking Rapid Erosion of the Mekong River Delta to Human Activities," *Scientific Reports* 5, no. 14745 (2015).

21. Hong Kong Government Records Service archives, record ID HKRS58-1-36-13.

22. Édouard Glissant, *Poetics of Relation,* trans. Betsy Wing (Ann Arbor: University of Michigan Press, 1997).
23. Kathryn Yusoff, *A Billion Black Anthropocenes or None* (Minneapolis: University of Minnesota Press, 2018).

Index

© The Editor(s) (if applicable) and The Author(s), under exclusive
license to Springer Nature Switzerland AG 2020
J. Carruthers and N. Dakkak (eds.), *Sandscapes*,
https://doi.org/10.1007/978-3-030-44780-9

Printed in the United States
By Bookmasters